Cells and Tissues

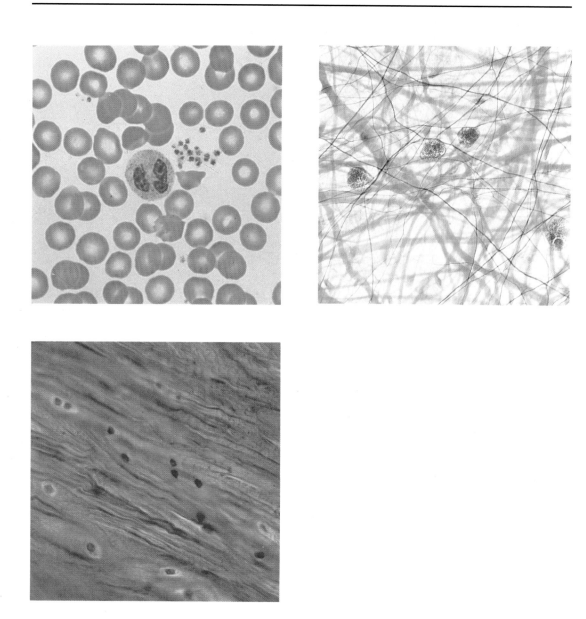

To identify these specimens turn to pages 66, 93 and 138.

Cells and Tissues

An Introduction to Histology and Cell Biology

Andrew W. Rogers

School of Medicine
The Flinders University
of South Australia

ACADEMIC PRESS 1983

A Subsidiary of Harcourt Brace Jovanovich, Publishers

London   New York
Paris   San Diego   San Francisco   São Paulo   Sydney   Tokyo   Toronto

ACADEMIC PRESS INC. (LONDON) LTD
24/28 Oval Road, London NW1 7DX

United States Edition published by
ACADEMIC PRESS INC.
111 Fifth Avenue, New York,
New York 10003

British Library Cataloguing
in Publication Data
Rogers, A W
  Cells and tissues

  1. Cells   2. Histology
  I. Title
  611'.018   QM551

  ISBN 0-12-593120-4

  LCCCN 82-074010

  Printed in Hong Kong
  by Imago Publishing Ltd

Preface

This book has grown out of the frustration of many classes of medical students, as they started the microscopy of cells and tissues. I have become persuaded that a small book, oriented towards helping them learn this new skill, could turn histology from a struggle into something interesting and useful. In particular, I acknowledge the students at Flinders and Oxford whose comments have directly helped in the production of this book. I shall look forward to their suggestions, as well as those of their successors, on how it may be improved.

I have received a great deal of help and encouragement from colleagues while preparing the book. In particular, my colleagues in the Human Morphology Unit — Drs C. Straznicky, J.B. Furness, B.J. Gannon, M.E. Jones and C.R. Murphy — have made many suggestions, spotted many errors and provided many pictures. Graduate students and assistants — Drs T. Williams, J. Bertram and Miss C. Lunam — have likewise contributed pictures. I am grateful to Drs J. Bradley and J.D. Henderson for commenting on drafts of particular chapters. I am particularly grateful to Professor C.G. Phillips and Dr Savile Bradbury, for allowing me full access to the teaching and demonstration material in the Department of Human Anatomy, Oxford: many of the colour pictures were taken there. Dr Bradbury also read the entire manuscript, offering many helpful suggestions, as well as contributing an electron micrograph. Dr Margaret Haynes took many electron micrographs expressly for this book, drawing on her collection of human material at the Adelaide Children's Hospital. Other micrographs have been contributed by Drs D. Haynes and J.D. Henderson, of the Histopathology Unit at Flinders; by Dr J. Fanning, of the Pathology Department of Adelaide University; by Dr J. Heath, of the School of Medicine, Newcastle, NSW; and by Mr C. Yeo, of the Anatomy Department, University of New South Wales.

Some of the sections for light micrographs were expressly prepared, with considerable skill, by Ms Michelle Lewis. Dr A. Morphett, of the Histopathology Unit at Flinders, lent me the slides of acute bronchopneumonia.

The typing has been a joint effort by Mrs B. Hammond, in Oxford, and Mrs S. Fiebig and Miss S. Lane in Adelaide.

Finally, the drawings were prepared in the Medical Illustration Unit at Flinders by Mr Alan Bentley.

To all of them, my heartfelt thanks.

Somerton Park, Adelaide   *A.W. Rogers*
October 1982

Contents

Chapter 4   Epithelia: the body's limits   42

Chapter 5   Connective tissue: the spaces in between   64

For Inta

1
Why study histology?

Histology is the meeting place of biochemistry, physiology and anatomy. Enzymes, nucleic acids, cyclic AMP and other active molecules do not float randomly in solution: we are not some sort of primeval soup held together in a test-tube of skin. These molecules are organized with great precision inside the cell into discrete structures, many of which can be recognized by electron microscopy. Cells are themselves arranged in intricate networks whose patterns in space are as necessary to the functioning of the body as the biochemical characteristics of the molecules out of which they are constructed. These patterns underlie and determine the shape and workings of limbs, brain and internal organs. We cannot understand the vertebrate body without taking into account this intermediate level of organization, without looking at populations of cells, the patterns they make, their functions and the control mechanisms which govern them. The processes of damage and repair, too, reflect the activities of these cell populations and their carefully regulated co-operation.

Within a complex and relatively large body such as ours, cells vary considerably in shape and function. Yet each one contains an identical set of genes. There must be great evolutionary advantage in being built out of many tiny cells, to outweigh the vast expenditure of energy involved in reduplicating the genetic information so many times. The process by which cells with the same genetic information come to differ considerably from each other, even when sharing a similar micro-environment, is called differentiation. It involves two parallel chains of events: the development of components and structures which are increasingly specialized, and the loss of options to develop in other ways. This process is outside the immediate scope of this book, since it would take us far into experimental embryology; it has been discussed most readably by Wessels (1977). This book will take the existence of differentiated cell types for granted, and study their interactions and behaviour in the body.

Any section through a tissue or organ contains many hundreds of cells. Yet it is possible with a little training to identify the organ involved. In fact it is much easier to name the organ than to say from which species it came. The fact that the cells making up an organ form the same, recognizable pattern, even in different species, implies a high degree of control over cell shape, size, orientation, spacing and activity. This precise regulation can be regarded as structural homeostasis. The co-operation of large numbers of cells of various types to produce such patterns must require communication between cells, so that each cell takes up a shape, orientation and distance from its neighbours that are appropriate to its position. We are surprisingly ignorant of how these factors are controlled: research into the organization of cells into complex tissues is gathering momentum and is likely to produce fascinating results in the next decade or so.

Sadly but inevitably, most students are severely practical about histology. It is usually seen as a large mass of facts to be learnt and sections to be identified. Given time, both feats can be achieved: it does not take much intelligence, only effort and a good memory. But the similarities between organs and tissues, or between electron micrographs of one cell type and another, reflect similarities in function, providing us with correlations that are often simple and direct. The search for recurring patterns and the effort to understand them are not just research tasks for a handful of cell biologists: they provide a way in which you can make the study of histology easier and more enjoyable, transferring some of the burden from your memory to your intelligence.

Most textbooks of histology are comprehensive, listing all the major features of each organ or tissue. They make excellent reference books, but their very design makes it often difficult for you to pick up the clear, recurring correlations between structure and function which can be made, until you have read most of the book. Inevitably, such books encourage the memorizing of facts and appearances. This book is quite selective: many organs do not even get a mention. Instead, it attempts to focus on common patterns and their interpretation. This book cannot replace the standard histology textbook; but, if you read this book carefully, fact after fact in the reference book will seem reasonable and right, fitting into the framework you will have acquired which links structure to function.

So read this book to enjoy it. It was written to be self-contained, not requiring a microscope and a set of slides. It was written for students starting in histology, though it does assume some elementary knowledge of biochemistry and of the structure of the mammalian body. At the end of each chapter is a suggested list of further reading for those who are interested in the topic; inevitably, this list is out-of-date by the time it reaches you. Most of these references are to review articles, and it will help to note the journals in which they are found, such as *Scientific American*, and to exercise your curiosity by glancing at recent issues.

Finally, I have found over the years that it is remarkably easy to read a page of print without retaining any of it to the next day. So I have placed a number of questions in the text, set in italic type. Each question can be answered from information already given. At the end of the book, there are brief notes on each question indicating the sort of answer expected and where the material is covered. This is a device to emphasize and repeat important material, and should help to make its retention easier.

If all that you want is to pass your exams in histology, please read this book early in your course. By introducing you to recurring patterns and discussing their interpretation, I hope to make your study of histology easier, more enjoyable and considerably more intelligent. I warn you that I will also try to convey to you some of the fascination of cells, which form a society of disciplined and ordered beauty so strange that it is difficult at times to believe one is watching populations within one's own body.

Further reading

Wessels, N.K. (1977). "Tissue Interactions and Development." W.A. Benjamin, Menlo Park, Calif. Easy to read and stimulating account of cell and tissue interactions in embryology. Read Chapters 1–3 for recent ideas on differentiation.

The techniques available

In a very obvious sense, the techniques that exist for studying cells and tissues determine our knowledge. The very existence of cells was unknown until lenses were developed which permitted them to be seen. Until the development of electron microscopy, the ultrastructure of the cell was a mystery. But the influence of techniques on our knowledge is more far-reaching than that. Histology grew into a science in the second half of the nineteenth century, a time which saw steady improvement in the performance of the light microscope, and also the appearance of a chemical industry which, particularly in Germany, was often directed to producing new textile dyes. Sections prepared then provided the basis for the descriptions and classifications of cells which we still use today. Each new combination of cell shape, size, orientation, position and staining characteristics became a named cell type. The relationships between various cell types were continually and often bitterly debated, because standard histological techniques did not permit evidence to be collected on transformations between one cell type and another.

So cells that look alike, such as lymphocytes, have a common name, though we now know that this "cell type" includes cells with at least two very different life histories. Cells that look different have different names, though we now know that, for instance, the B-lymphocyte and the plasma cell are different stages in the life of the same cell. All through histology run references to "basophilic" (loving basic dyes), "acidophilic" (loving acidic dyes), "chromophobe" (not staining with common dyes) and "metachromatic" (staining, but having a colour that is different from that of the original dye).

Cells can be described and classified in many different ways which are valid and useful. The classification we use grew out of the techniques available when the detailed study of tissues and organs began.

Dimensions

The fertilized ovum is amongst the bigger mammalian cells, ranging from 80 to 140 μm in diameter, depending on species. (10^6 μm are 1 metre.) It is just visible with the unaided eye, under good conditions of illumination. The majority of mammalian cells are 7–20 μm diameter; the red blood cell is a useful index of size in sections, and is 7·2 μm diameter, in fixed material.

Cell nuclei are often in the size range of 7–12 μm diameter. Mitochondria, the cell organelles associated with the production of cellular energy by oxidative phosphorylation, vary somewhat in size and shape: an average one would be 1–2 μm long and about

0·2 μm diameter. The limit of resolution of the light microscope is about 0·2 μm, so mito-chondria can be seen in suitable preparations. Ribosomes, the sites of protein synthesis, are about 15 nm diameter (10^9 nm are 1 metre), and require electron microscopy for their visualization.

Clearly, we need the light microscope to study populations of cells adequately, and electron microscopy if we wish to examine the structures within the cell.

The light microscope

The development of the compound microscope was a gradual process involving scientists from many European countries, but, by the latter part of the nineteenth century, instruments were available which permitted microscopy under conditions not so different from those we use today. I have one of the early factory-made instruments, a Leitz from 1896, and I still use it profitably. The history of the microscope has been well described by Bradbury (1974).

The design of the microscope

The light microscope can be conveniently considered in two parts: that between the source of light and the specimen, and that between the specimen and the viewer's eye.

Light consists of photons, which move in a periodic path like a sine wave. A single, narrow beam of white light contains photons of wavelengths corresponding to all the colours of the rainbow. The photons of any one wavelength are not in step, but randomly moving, out of phase with each other. The plane in which the photons oscillate also varies randomly, with no agreed plane of vibration. Such a pencil of light, when it passes into a different medium of higher refractive index, such as from air to glass, behaves rather like a car passing from a properly surfaced road on to loose sand: if it hits the boundary squarely, at right angles, it continues in a straight line at lower speed. If, however, it hits the boundary at an acute angle, the front wheel that hits first becomes slowed while the other one keeps going, turning the car as it enters the new surface. A similar turn, but in the opposite direction, occurs when the car emerges back on to a smooth, hard surface.

If a layer of glass is parallel-sided, the pencil of light will emerge on the other side still travelling in the original direction. If, however, the glass is a biconvex lens, perfectly symmetrical in shape, the pencil of light will be permanently redirected, unless it happens to pass through the very centre of the lens where the two surfaces are parallel. Such a lens has the property of making parallel rays of light converge to a point; the distance between the lens and the plane on which such points lie is called the focal distance of that lens (Fig. 2.1a). A point source of light on the focal plane will obviously give rise to parallel rays of light emerging from the lens. If the point source lies somewhere between the focal plane and infinity, the lens will produce a real image lying between infinity and the focal plane on the opposite side of the lens (Fig. 2.1b). If, however, the point source is nearer to the lens than the focal plane, the light rays emerging from the lens will be divergent, and no real image will result; an eye receiving these rays will interpret them as coming from a virtual image on the same side of the lens as the source of light (Fig. 2.1c).

The microscope produces two stages of magnification between the specimen and the eye (Fig. 2.2). The first is a real image of the specimen, produced by the objective lens. The eyepiece then produces a virtual image of this first, real image.

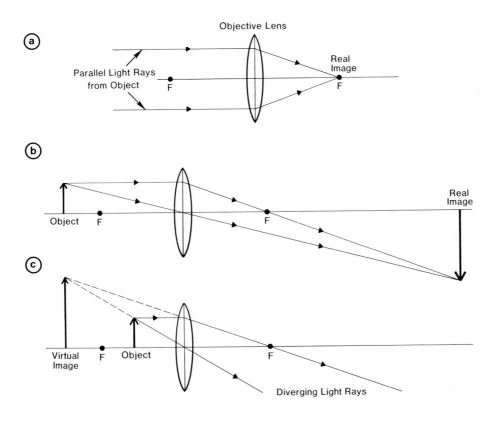

Fig. 2.1   Light paths through a biconvex lens. F is the intersection of the focal plane with the optical axis.

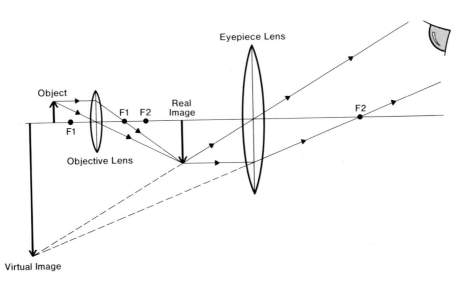

Fig. 2.2   The light path from object to eye in a compound microscope. F1 and F2 are focal points of objective and eyepiece lenses respectively.

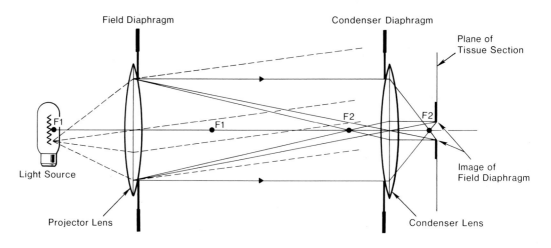

Fig. 2.3   The light path from light source to specimen in a compound microscope. F1 and F2 are focal points of projector and condenser lenses respectively.

The microscope thus takes the light rays falling on a small area of the specimen and spreads them over a much larger area of virtual image. Clearly, the virtual image will be much fainter than the light intensity at the specimen. The optical path between the light source and the specimen (Fig. 2.3) is designed to focus light on the specimen so that the final, virtual image is bright enough to see. It has two other functions: to reduce glare, or stray light, passing through the eyepiece which does not contribute to the formation of the final image, and to illuminate the specimen evenly.

Even illumination can be achieved in several ways. The simplest is to use an opal bulb as the light source. The more common way in today's microscopes is to place a projection lens in front of the light source, so that the light is at the lens' focal plane. Light from each point on the source, which is usually the filament in an electric bulb, emerges from the entire surface of the projector lens as a parallel series of rays; this is illustrated by the dashed lines on Fig. 2.3. In effect, all the points on the surface of the projector lens become point sources of light of equal brightness, and we can consider the projector lens as the source of illumination.

The condenser lens collects light from this source and brings it to a focus in the plane of the specimen; the solid lines indicate these rays on Fig. 2.3. Thus the light intensity at the specimen, and hence at the final image, is greatly increased.

Two diaphragms exist in the light path to the specimen. One, near the projector lens, is called the field iris. This is adjusted to limit the diameter of the illuminating beam so that it just fills the field of view at the eyepiece. A larger illuminating beam produces glare. The second diaphragm, the condenser iris, matches the cone of light entering the specimen to that which can be collected by the objective lens. This is further discussed on p. 7.

Magnification and resolution

Magnification is the size of the final image compared to that of the specimen. It is possible to produce any desired magnification: you can take a photograph down the microscope and enlarge it to the size of a football field if you wish.

Resolution is the ability to resolve into separate images two structures that are very close together in the specimen. Resolution depends not only on the magnification of the lens system, but on the amount of glare and contrast in the final image. Resolution is finally limited by the wave motion of photons. However good the lenses and however high the magnification, points lying less than $0 \cdot 2$ μm apart in the specimen cannot be resolved by normal light microscopy. The highest useful magnification is about $\times$ 2000. Above this, no further increase in resolution can be obtained: it is empty magnification.

If we think of photons leaving a point in the specimen, only a given percentage of them will be collected by the objective lens, the rest passing outside the perimeter of the lens. If the lens is larger in diameter or nearer to the specimen, this percentage collected will be higher, the final image will be brighter, and more of the information leaving the specimen will have gone into image formation. The numerical aperture is a measure of the fraction of light leaving the specimen that is collected by the objective lens. The best objective lenses have a high numerical aperture ($1 \cdot 3$ or over), and, since it is inconvenient to get this by means of a very large lens, such lenses are placed a very short distance from the specimen.

One final point about glare. Light is scattered to some extent every time it passes from material of one refractive index to material of another — from air to glass, for example. Where the highest resolution is needed, this source of glare is reduced by placing a drop of oil between specimen and objective lens: the oil has the same refractive index as glass. Oil immersion objectives are those giving the highest magnification (usually about $\times$ 100), and having the highest numerical aperture.

The use of the light microscope

The detailed steps of setting up a microscope for use vary a little from model to model. Some principles remain constant, however.

The condenser should focus the source of light accurately on the specimen. But the light source is effectively the front of the projection lens.
Where is this lens? What is its function? (Note 2.A)
This cannot easily be seen or focused. Instead, the field iris, lying in front of it, is focused. With the specimen in focus at the eyepiece, the field iris is closed right down and its edges brought into focus by moving the condenser lens up and down. When it is in focus, it should be centred accurately, then opened until its edges just disappear from the field of view.
What will happen if the field iris is opened more widely? (Note 2.B)
Next, the condenser iris should be adjusted. Open it fully and remove an eyepiece. You will see an illuminated circle down the tube of the microscope. Now close the condenser iris until the diameter of this circle is reduced by a quarter to a third. Replace the eyepiece.
What will happen if this iris is too widely open? (Note 2.C)
These simple adjustments get the best out of your microscope, and should become a habit.
Which, if any, of these adjustments will need to be made again if you change to another objective lens? Why? (Note 2.D)

The preparation of tissues for microscopy

Some cells, such as those circulating in the blood, are already separate, and only require spreading thinly on a slide to be ready for viewing. Most of our tissues and organs, however, are too solid for this, and need cutting into sections thin enough to allow light to penetrate. An image can then be formed without confusion from cells overlying other cells.

Fixation, embedding and sectioning

Even cells that are already in suspension, such as blood cells, will change in shape as they dry out and will decompose in time, particularly if they contain digestive enzymes. Somehow, the shape and size of cells must be made permanent. This is achieved by chemical fixation, a process of denaturing proteins and cross-linking them to provide a solid structure in place of the delicate membranes and interacting molecules of the living cell. The most commonly used fixatives are formaldehyde and glutaraldehyde, the latter being the more effective as a cross-linking agent.

Even after fixation, tissues are too soft to be cut easily into very fine sections. The standard technique of sectioning involves impregnating the tissue with paraffin wax and then sectioning the wax block. But paraffin wax does not mix with water. After fixation, all the water in the tissue is replaced with increasing concentrations of alcohol; the final absolute alcohol is then replaced with a non-polar solvent, such as xylene or toluene, in

Table 1 A Typical Schedule for Preparing Histological Sections

Fixation	10% formalin in phosphate buffer (pH 7·4)	24–48 h
	Running water	2 h
Dehydration	70% ethyl alcohol	18 h
	80, 90 100 and 100% ethyl alcohol	2 h each
Embedding	Chloroform	18 h
	Molten paraffin wax (50°C) × 3	2, 2 and 1 h
	Cast block in paraffin wax	
Sectioning	Cut sections at 5 μm	
	Float sections on water on slides to remove wrinkles, then dry	
Hydration	Xylene	2 min
	100, 90 and 70% ethyl alcohol	2 min each
	Several rinses distilled water	2 min
Staining	Haematoxylin	10 min
	Distilled water	rinse
	1% hydrochloric acid in 70% ethyl alcohol	5–10 s
	0·02% lithium carbonate in water	2 min
	Distilled water	rinse
	1% eosin in 90% alcohol	1 min
Mounting	100% ethyl alcohol × 3	1 min each
	Xylene × 2	2 min each
	Cover with mounting medium and coverslip	

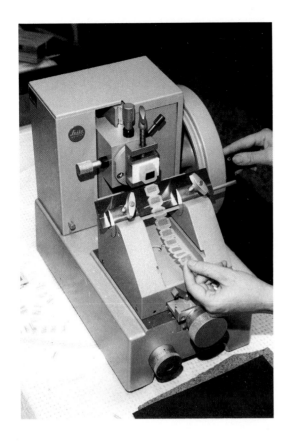

Fig. 2.4 Sections being cut from a paraffin-embedded block of tissue.

which the wax is freely soluble. The block of tissue is then soaked in paraffin wax at a temperature above its melting point, the wax is allowed to set on cooling, and the block of wax including the tissue is ready for sectioning.

This routine, which is summarized in Table 1, produces a block that can readily be sectioned down to a section thickness of 2–3 μm. It is reasonably simple and convenient. It does, however, remove many components from the tissue. Clearly, the fat solvents such as alcohol and xylene will dissolve away any fat in the section, while the chemical fixation and subsequent washing, together with the lower concentrations of alcohol, remove all the ions and many small molecules. Paraffin wax has another inconvenience, in that it is opaque and has to be removed before staining the section. Table 1 lists the further sequences of xylene and alcohol needed to bring the section to a stage at which water-soluble dyes can be applied to it, and then to remove the water once again to make a permanent, bacterium-proof specimen with the same refractive index as glass.

Why is this refractive index helpful to high resolution microscopy? (Note 2.E)

Sectioning is carried out on a microtome (Fig. 2.4), a sort of specialized bacon-slicer, and section thicknesses usually range from 3 to 12 μm with paraffin wax. If thinner sections are wanted, the tissue may be embedded in araldite or some similar polymer instead of wax. This gives a much harder block, from which sections can be cut with a steel or glass knife at thicknesses from 0·5 to 2 μm. The plastic is transparent, and already has a refractive index near to that of glass, so it may be left in place after sectioning; but, in this case, stains must penetrate this plastic in order to produce contrast in the section, so removal of the plastic is often carried out to simplify staining.

Staining the section

In routine histology and histopathology, the most widely used stains are haematoxylin and eosin, applied to the section one after the other. Haematoxylin is a purple-blue dye which in effect stains acidic structures in the section. After the lengthy process summarized in Table 1, the major acidic groups remaining in the section are complexes of DNA and RNA with protein. Cell nuclei contain the genetic DNA together with much RNA, so nuclei stain with haematoxylin. In addition, cells which contain an abundance of ribosomes and are rich in RNA will have a bluish-purple tinge to the cytoplasm. Eosin is a very non-specific stain which colours most proteins, particularly basic ones, a pinkish red. Such staining is shown in Fig. 2.5. Note that all the cell nuclei are clearly picked out by haematoxylin, and that, in addition, the cytoplasm of some cells is not pale pink, but rather purplish: this is called cytoplasmic basophilia, and indicates a cytoplasm rich in ribosomes.

Eosin does not distinguish between cytoplasmic proteins and those of the extracellular spaces, such as collagen. Many staining routines have been devised to overcome this problem. Haematoxylin usually stains the nuclei, whereas other dyes differentiate collagen from cytoplasmic proteins. There are many such trichrome stains, one of which is illustrated in Fig. 2.6; be prepared for many vivid colours and remember that in common use they distinguish collagen from cytoplasm.

The same distinction can be achieved by haematoxylin and Van Gieson's stain, which colours nuclei purple-blue, cytoplasm a yellow-green and collagen red (Fig. 2.7); this stain is often combined with a dye that specifically stains elastic fibres black, Verhoeff's stain.

Complex mixtures of dyes which are suitable for staining paraffin sections of tissues seldom penetrate well enough into plastic sections. These are often stained with a single dye — toluidine blue or methylene blue — which gives sufficient colour to the various major tissue components to allow them to be seen (Fig. 2.8).

Staining is much more complex than this brief summary implies. Those interested in following up its intricacies are referred to Baker (1960).

The staining techniques illustrated in this book are summarized in Table 2.

The limitations of conventional techniques of microscopy

The standard methods of histology, summarized in Table 1, are simple and reproducible. They produce a stained section which is thin enough for detailed observation, permanent and mounted in a medium with the same refractive index as glass. The sections retain most of the information we need on cell size, shape, orientation and spacing and tell us something of the extracellular structures present. But the methods used impose severe limitations on our microscopy. Specialized techniques of microscopy have in general

Fig. 2.5   (above, left) A section of human pancreas, stained with H & E. At this magnification, the nuclei of individual cells appear as small, dark specks: the diameter of a red blood cell would be 0·72 mm. (× 100.)
Fig. 2.6   (above, right) Human pancreas stained by Masson's trichrome technique. (× 100.)
Fig. 2.7   (below, left) Human pancreas stained by H & VG. (× 100.)
Fig. 2.8   (below, right) A plastic-embedded section of monkey kidney stained with toluidine blue. The arrow indicates red blood cells in capillaries of a glomerulus. (× 600.)

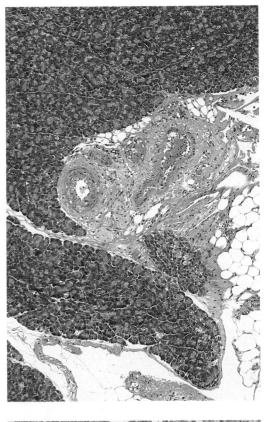

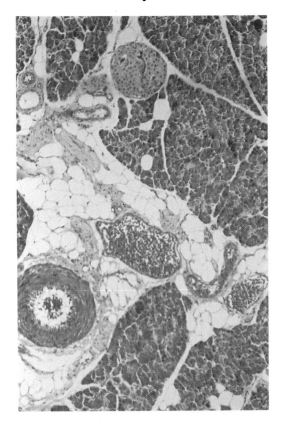

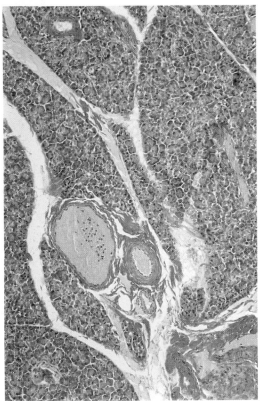

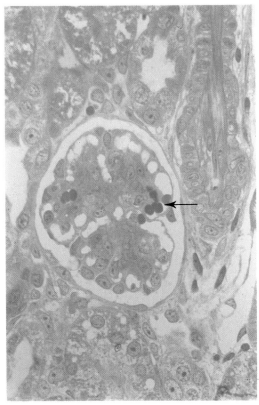

Table 2 Some Common Histological Staining Methods

Stain	Chief use	Results
Haematoxylin and eosin (H & E)	General tissue stain	Nuclei — blue or purple Basophilic cytoplasm and cartilage — purple Cytoplasm and collagen — pink RBCs, eosinophil granules — bright red

(Haematoxylin may be used alone, staining nuclei blue-purple. Variants exist, such as iron haematoxylin, which gives a black to grey staining of nuclei).

Stain	Chief use	Results
Haematoxylin and Van Gieson's stain (H & VG)	General tissue stain	Nuclei — brown-black Cytoplasm — yellow Collagen — red
Verhoeff's and Van Gieseon's stain (Verhoeff's & VG)	General tissue stain	Nuclei — brown-black Cytoplasm — yellow Collagen — red Elastic fibres — black
Mallory's trichrome	General tissue stain	Nuclei — blue-black Cytoplasm — red to mauve RBCs — bright orange Collagen — blue
Masson's trichrome	General tissue stain	Nuclei — black Cytoplasm — red to mauve Collagen and mucus — green

(N.B. There are many variants of the trichrome stains, giving slightly differing results)

Stain	Chief use	Results
Toluidine blue	Plastic-embedded sections	All tissues — shades of blue Mast cell granules — red-purple
Carmine	General stain, especially embryonic tissues	Nuclei — red Other components — lighter shades of red
Azocarmine (Azan)	General tissue stain	Nuclei — red Cytoplasm — lighter red Collagen — blue
Phosphotungstic acid/Haematoxylin (PTAH)	General tissue stain	Nuclei and muscle striations — blue Collagen — brownish red
Orcein and light green	Connective tissue	Nuclei — faint green Collagen — green Elastic fibres — black or brown Mast cell granules — purple

(Orcein may be used without the light green counterstain, giving brown or black elastic fibres)

Table 2 (Cont'd)

Stain	Chief use	Results
Leishman's stain	Blood smears	Chromatin — purple Basophilic cytoplasm — pale blue Neutrophil granules — pink to purple Eosinophil granules — red Basophil granules — dark purple
PAS method	Carbohydrates	Nuclei — blue Mucus, GAGs and glycogen — magenta
Pyronin and methyl green (PMG)	Nucleic acids	Chromatin (polymerized DNA) — green or purple Nucleoli and cytoplasmic RNA — red
Cresyl fast violet	Nervous tissue	Nissl substance — purple
Koelle–Gomori method for cholinesterases	Nervous tissue	Sites of cholinesterase activity — dark brown
Oil red O and haematoxylin	Lipids	Nuclei — blue Lipids — bright red
Osmic acid	Lipids, Nervous tissue	Lipids, including myelin — black
Silver impregnation	Nervous tissue Reticular fibres	Neurones and reticular fibres — black

(Many variants exist with slightly differing effects: these include Holmes' and Bielschowsky's methods)

evolved to answer problems that cannot be tackled by the observation of routine sections, so we will consider them under the headings of the various major drawbacks of conventional histology.

The tissue is dead

Fixation has killed the cells and denatured their proteins. Cell division, cell migration and cell activity all have to be inferred from a dead, static specimen.

It is possible to view living cells in the light microscope. The standard staining method of Table 1 introduces contrast by colouring the tissue, which is viewed by white light. If, however, all the light reaching the specimen from the substage condenser is of the same wavelength and, in addition, all the photons are oscillating in phase, then differences in refractive index within the specimen may slow or speed up photons that come through different areas of the cell or tissue, and these differences in phase of the light emerging from the specimen can be converted into a black-and-white picture. This is, very simply, the principle of phase contrast microscopy and its associated techniques, interference and interference contrast microscopy. These methods can be used to study living cells and

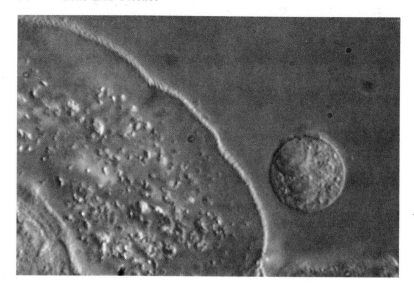

Fig. 2.9 Live cells viewed unstained by interference contrast. At left, part of a flattened epithelial cell from the lining of the mouth: the small, round cell (centre) is a polymorphonuclear leucocyte. (× 1480.)

tissues, unstained, provided they can be made thin enough (Fig. 2.9): ideally, monolayers of cells are needed.

Other tricks are made possible by controlling the light falling on the specimen. If the plane of oscillation of all the photons is identical, the light is said to be polarized, and any structure in the specimen that interferes with this polarization can be visualized. More simply still, the specimen can be viewed by white light converging on it at a considerable angle from the optical axis. Most of this light will pass through the specimen and never enter the objective, but some structures and membranes will scatter and reflect light into the objective, appearing bright against a black background. This dark ground microscopy can also be used to view living cells.

The stains are non-specific

Instead of colouring all nuclei blue and everything else pink, can we carry out controlled chemical reactions on sections to introduce colour into one particular chemical? In other words, can we study the distribution of, for instance, glycogen, or DNA, or particular enzymes in a tissue section?

Histochemistry is the science of detecting specific chemical groups in tissues; cytochemistry detects them at the level of individual cells. Early histochemical tests relied on known chemical reactions that led to a coloured end-product, and applied these to tissue sections. Soon, a veritable new industry grew up, devising and validating new reactions for use in microscopy. The range of histochemical tests now available covers every class of compound and goes far beyond titration against a coloured reagent. Enzyme histochemistry, for instance, involves incubating the section in a medium containing a substrate for the enzyme concerned: one portion of the substrate molecule, after being split by the enzyme, is captured and precipitated by some component of the medium, and this precipitate indicates the distribution of the enzyme in the tissue. Figure 2.10 shows the distribution of the enzyme, acetylcholinesterase, on the surface of a single, microdissected muscle fibre.

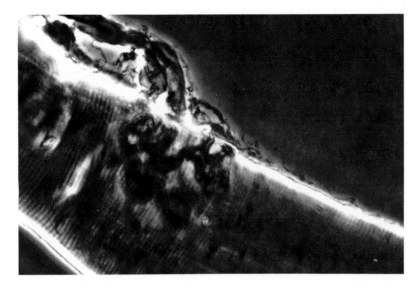

Fig. 2.10 A single, micro-dissected fibre of striated muscle from a mouse. The nerve (above) ends in a darkly stained series of curved furrows on the surface of the fibre. Koelle-Gomori method for cholinesterases. (× 410.)

A histochemical reaction for glycoproteins is shown in Fig. 4.10a, and one for nucleic acids in Figs 8.6 and 8.7. A fluorescence technique for catecholamines is shown in Fig. 12.8.

Antibodies are the most sensitive tools we possess for identifying specific proteins. If antibodies carrying a fluorescent marker are applied to a section, the sites where fluorescence is seen are, in well-controlled experiments, those where the specific protein used as an antigen occurs.

It is possible to study the distribution of almost any compound in the tissues by histochemistry, and often to measure the amount present in particular sites.

Few chemicals survive the preparative techniques

It is no good devising a chemical reaction to identify a particular compound in tissues if that compound is removed by histological processing, or denatured to such an extent that it becomes unreactive. Fortunately, a technique exists to section fresh tissues, without fixation or embedding. This involves freezing the fresh tissue, effectively converting its contained water into an embedding medium of sufficient hardness to permit sectioning. Many sequences of histochemistry become possible with frozen sections, with or without fixation. If frozen sections are so good, why do we continue to use the nineteenth century's paraffin wax? It is technically much more difficult to cut frozen sections, and, at high resolution, ice crystals which formed on freezing can be seen to have produced many small, irregular tears in the tissue.

A section is only two-dimensional

Tissues and organs are three-dimensional, and it is not easy to visualize their organization from two-dimensional sections.

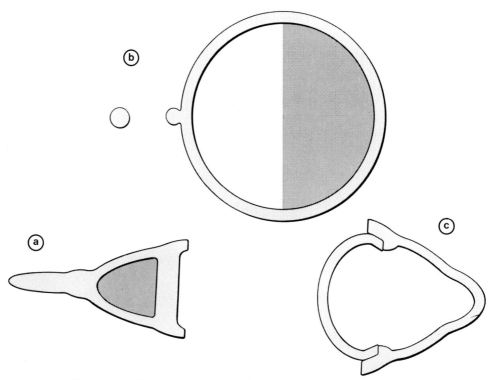

Fig. 2.11   Three parallel sections through a common household object.

Figure 2.11 shows a series of sections through a very familiar object. Can you identify it? (Note 2.F)
Often, we cannot use very thick sections, as they will transmit too little light, and the superimposing of many layers of cells will produce too confusing a picture. It is possible to cut and examine a long series of adjacent sections, and reconstruct the three-dimensional shape of structures from them. This is very time-consuming, but has been the basis of a great deal of embryological histology.

Stereology is the science of predicting and measuring three-dimensional structure from two-dimensional sections, and it has reached quite a degree of sophistication. Morphometry is the measurement of shape, and it overlaps with stereology: indeed many histologists use the terms interchangeably. These are groups of statistical techniques devised to overcome the two-dimensional nature of the histological section.

Of course, the surfaces of specimens can be viewed by dark ground microscopy and their shapes can be detected directly, provided the specimens are firm enough.
What is dark ground microscopy? (Note 2.G)
Unfortunately, many of the structures we are interested in are buried in a continuous, three-dimensional matrix composed of other cells and extracellular material, and cannot be separated out for surface microscopy.

Holography, or the generation of a three-dimensional microscopic image, is possible at fairly low magnifications, but is little more than a party trick at the moment.

The section is static and unchanging

It is possible to analyse the cells in a section, recording every detail, but none of these data

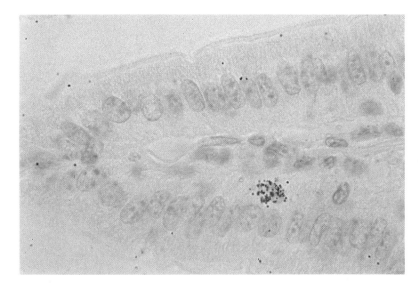

Fig. 2.12 Autoradiograph of small intestine from a mouse killed 1 h after an injection of ³H-thymidine. One nucleus only is labelled. Haematoxylin. (× 300.)

tell us where the same cells were 5 minutes or 2 days before the tissue was fixed, or what they looked like then. Living tissue undergoes amazing transitions: big cells divide to give small cells; long, thin cells round up and grow large; cells change their orientation and wander off on long migrations. None of these transformations can adequately be studied from fixed material, and very few of them can be reproduced in tissue culture conditions suitable for the microscopy of living cells. It is like trying to understand traffic flow in a city from a single aerial photograph.

The tracer experiment provides the answer to this problem. If cells or molecules marked in a recognizable but non-damaging way are introduced into the tissue, their distribution and appearance can be found at later times, and some attempt be made to reconstruct their activities. Radioactive isotopes are the most widely used method of labelling cells or molecules. The radioactivity can be detected and given a position in the section by means of autoradiography, a technique by which the section is covered with a thin layer of a specially produced variant of photographic emulsion — nuclear emulsion. The radioactive particles leaving the section make developable the crystals of silver halide which they hit, so that silver grains lie over those regions of the section which are radioactive (Fig. 2.12).

Other labels can be used. Cells with recognizable "marker" chromosomes can be introduced into the body, and so on, but autoradiography is the most versatile and widely used tracer technique. Autoradiography has other uses: it can be used as a straightforward histochemical technique, identifying the sites of binding of a radioactive reagent, for instance.

How can one label cells? Most of the materials out of which cells are made are being continually recycled, and would not provide suitable labels. Only one cell component is really suitable. The DNA, the genetic material in the nucleus, remains virtually unchanged through the lifetime of a cell: it becomes doubled in the cell as new DNA is formed before cell division, the mixture of new and old DNA being distributed between the two daughter cells. Labelling the DNA of a cell, at the division which gave it birth, by allowing it to incorporate thymidine labelled with tritium, is the most widely used method for tracing that cell's migrations and changes in appearance.

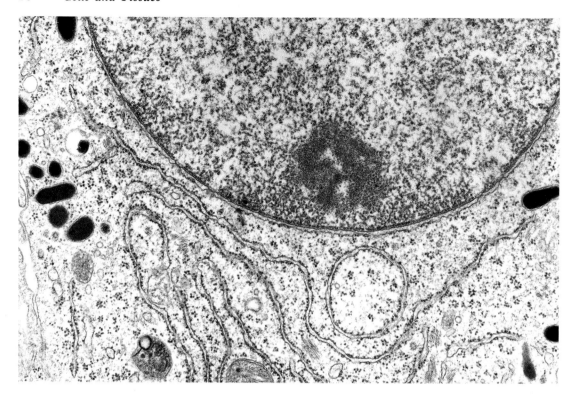

Fig. 2.13 Transmission electron micrograph (TEM) of part of a single cell. Part of nucleus (above): the cytoplasm contains many clusters of black dots, which are ribosomes. (× 30 000.)

The resolution is limited

We cannot observe structures below the limit of resolution of the light microscope. *What is the limit of resolution? What factor is responsible for imposing this limit? (Note 2.H)* The transmission electron microscope (TEM) was first developed by Knoll and Ruska in 1931. A stream of electrons is directed at the specimen, instead of a beam of light. Electrons are easily scattered on passing through matter, including air, so it is necessary for the electron beam to travel through a high vacuum: in this way, only the specimen produces scattering. Electrons travel with a wave motion similar to that of photons, but with far shorter wavelengths, so that far higher resolutions can be achieved in the TEM. Electrons are charged particles, so that their paths can be influenced by magnetic fields: the lenses of a TEM are electromagnets, and different magnifications are achieved by varying the current flowing through the lenses. Electrons cannot be detected by the eye, so, after passing through the column they are made to hit either a phosphor, which emits photons in proportion to the density of electrons hitting it at any given point, or a photographic plate. The electron micrograph (Fig. 2.13) is a shadow of the specimen, the light areas indicating places where the electron beam reached the plate without significant scattering, the dark areas showing the presence of material which scattered the electrons. Material of high atomic number has greater scattering power than lighter elements.

Given these basic differences between the light and electron microscopes, the fundamental design of the two is very similar (Fig. 2.14). An electron gun replaces the light source, producing a stream of electrons at a steady accelerating voltage. This illuminating beam is directed on to the specimen by a condenser lens. After passing through the specimen, the emerging electrons are spread by an objective lens, producing magnification. A virtual image, such as that produced by the eyepiece of the light microscope, is useless since it will not be viewed directly by eye: a real image on the phosphor screen or photographic plate is needed, and this is produced by the projector lens.

The techniques of preparing specimens for the TEM are similar to those for light microscopy. Fixation, embedding in plastic, sectioning (usually at thicknesses around 100 nm) and staining follow in sequence. Staining involves introducing atoms of high atomic weight: uranyl acetate and lead salts are common stains. The sections are not mounted on a glass slide, but on a fine grid of copper or platinum, and the areas of tissue between the grid bars studied.

The limit of resolution of the TEM is about 0·2 nm, given an ideal, very thin specimen. The resolution achieved with a plastic section is considerably worse than this, often around 1·2 nm.

The TEM has all the drawbacks of the light microscope, which have been listed over the last few pages. The section is dead, and each cell can only be looked at once; the stains are non-specific; the image is two-dimensional, and so on. Many special techniques have been devised to overcome these drawbacks: autoradiography and cytochemistry are examples, essentially similar to the same techniques at the light microscope level; stereology and morphometry are important tools in electron microscopy. Frozen sections have not been widely used in the TEM, largely because of the artefacts due to ice crystals, while the microscopy of living material is not so far possible, because of the high vacuum and the high radiation dose.

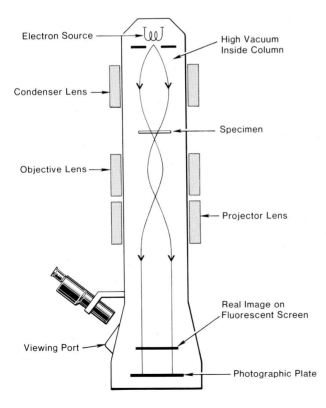

Electron Source

High Vacuum Inside Column

Condenser Lens

Specimen

Objective Lens

Projector Lens

Real Image on Fluorescent Screen

Viewing Port

Photographic Plate

Fig. 2.14 The basic design of a transmission electron microscope.

Fig. 2.15   A scanning electron micrograph of a plastic cast of the interior of a small arteriole (right) at its origin from an artery: retina of cat. (× 2000.) (Courtesy, Drs B.J. Gannon and G. Gole.)

The scanning electron microscope (SEM) is essentially a dark ground microscope, looking at the electrons scattered back towards the detector by a solid specimen (Fig. 2.15). It is remarkable for its depth of focus.

The electron microscope has one facility which is not technically possible in light microscopy. As electrons pass near an atomic nucleus, X-rays are generated whose wavelengths are specific for the element concerned. By analysing the spectrum of X-rays produced from a small part of the specimen, it is possible to identify the elements present and even measure their relative amounts. X-ray analysis is a cytochemical technique with great possibilities.

Collecting information is difficult

Nearly all the techniques we have discussed produce an image that ends up being looked at. Now the human eye and brain are notoriously bad at making measurements, and questions like "Are these nuclei larger than those?" or "Is this stain more intense here than there?" are very difficult to answer exactly by looking and estimating. A wide and increasing range of automatic and semi-automatic devices exists to make precise measurements on microscopic images. These range from machines that measure optical density to fully automatic image analysers capable, for example, of identifying, measuring and sizing cells of particular shape or optical density (Bradbury, 1976).

Back to earth

We have come a long way from your light microscope, and the section stained with haematoxylin and eosin. It is important for you to realize how little you can learn about living populations of cells by looking at such sections through such a microscope. It is just as important to know that techniques exist to extend greatly the range of information we can collect, to help us understand how these populations function and are organized.

Further reading

Baker, J.R. (1960). "Principles of Biological Microtechnique." Methuen, London. The early chapters contain an excellent discussion of the theoretical basis of histological staining methods.

Bradbury, S. (1974). "Microscope." Encyclopaedia Britannica, 15th Ed. Macropaedia, Vol. 12, 127–138. Encyclopaedia Britannica, Chicago. A clear and concise history of microscopy, and descriptions of light and electron microscopes.

Bradbury, S. (1976). "The Optical Microscope in Biology". Edward Arnold, London. A slim monograph describing the light microscope, and its special applications such as phase contrast, together with stereology and image analysis.

Everhart, T.E. and Hayes, T.L. (1972). The scanning electron microscope. *Scient. Am.* **226:1**, 54–69. The principles and construction explained.

Pearse, A.G.E. (1980). "Histochemistry, Theoretical and Applied", 4th Ed. Vol. 1. Churchill Livingstone, Edinburgh. The first chapter introduces the subject and gives an historical account of its development.

Rogers, A.W. (1979). "Techniques of Autoradiography", 3rd Ed. Elsevier, Amsterdam. Read the first chapter to see the scope of the method and the background to its development.

Weakley, B.S. (1981). "A Beginner's Handbook in Biological Transmission Electron Microscopy", 2nd Ed. Churchill Livingstone, Edinburgh. Chapter 1 gives a simplified account of the basic theory of electron microscopy.

3

The anatomy of the cell

Several processes tend to produce uniformity or homogeneity in a given volume of water that contains a number of dissolved compounds — diffusion, flow of water from convection and from movement of the container, and so on. The packaging of the same volume of water in a large number of individual cells allows the existence of differences in composition between one part of the whole and its neighbours, and interactions between these micropackages form the basis of processes much more complicated than those that are possible in simple solution. Cells in complex organisms form many, varied populations, providing immense flexibility and the possibility of recovery from injury. The energy invested in reduplicating the body's genetic material to provide a copy for each cell is amply repaid in the range of activities that multicellular structure makes possible.

The basis of cellular structure is the restriction of diffusion. The effects of this force, which tends to abolish differences between one part of a volume of water and the rest, are reduced by the cell membrane.

The cell membrane

Phospholipids are composed at one end of fatty acids, which are insoluble in water and tend to separate out from it, and at the other, of water-soluble phosphate groups. When a few phospholipid molecules are placed in water, the lipid ends of the molecules tend to aggregate, forming a micelle, with the phosphate groups projecting out into the water. Addition of more phospholipid molecules produces a sheet of phospholipid, two molecules thick, with the lipid ends of the molecules in contact, tail-to-tail, while the phosphate groups form heads out into the water. This pattern produces a very thin layer of lipid, running through the water, a layer that tends to form spheres or bubbles rather than flat sheets with free edges. Such a lipid bilayer is a significant barrier to the free diffusion of dissolved molecules and ions. The cell membrane has this general construction (Fig. 3.1).

But walls need doors. A drop of water isolated by a phospholipid bilayer from the aqueous solution surrounding it cannot exchange molecules freely with its environment or respond to it. To be effective, a cell membrane must permit, or even assist, chosen ions and molecules to cross the phospholipid layer. Biological membranes have particles, mainly protein, embedded in them, extending through from one side to the other: these are intramembranous particles (IMPs). These proteins appear to be held in the membrane by a central part of the molecule, which is hydrophobic and hence seeks out and stays in the

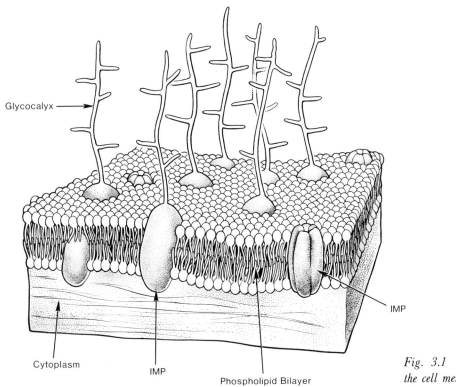

Glycocalyx

IMP

Cytoplasm

IMP

Phospholipid Bilayer

Fig. 3.1 The structure of the cell membrane.

lipid core of the membrane. Some of these particles form channels, permitting the diffusion of particular ions; others are enzymes, capable of transporting specific substances across the membrane. An example of the latter is Na–K ATPase, the enzyme which ejects sodium ions from the cytoplasm into the extracellular spaces, in return for potassium ions. As a consequence of its activity, the volume of water inside the cell contains high concentrations of ionic potassium and low levels of sodium, while the extracellular fluid is high in sodium and low in potassium. The creation of these differences is due to the activity of enzyme molecules embedded in the cell membrane; its maintenance results from the barrier to diffusion provided by the phospholipid bilayer.

Transmission electron microscopy of thin sections stained with osmium textroxide shows membranes as a dark line, resolved at high magnification into two parallel dark lines about 2·5 nm thick, separated by a clear line about 3 nm thick. It has been suggested that the central, clear line is the lipid core of the membrane, and the two parallel dark lines represent the phosphate groups and protein molecules associated with them, stained with osmium.

Freeze fracture

But the TEM is a clumsy instrument with which to look at membranes. Since membranes very rarely lie in the plane of the section, one tends only to see a thin slice of membrane in any one place. A technique known as freeze fracture can demonstrate large areas of membrane on a single electron micrograph. A small block of tissue is frozen solid and then struck smartly with a metal knife. This does not cut a section through the ice, but breaks it, usually into two pieces. The thin layers of lipid in the membranes are weak planes in the

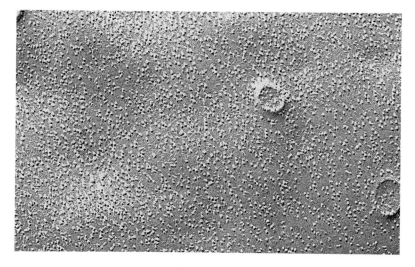

Fig. 3.2 Freeze fracture preparation of the apical membrane of a cell lining the cavity of the rat uterus, showing IMPs. The two round profiles are microvilli, broken off at the plane of the membrane. (× 76 500.) (Courtesy, Dr C.R. Murphy.)

ice block, and the preferred plane of fracture runs down the middle of the phospholipid bilayer, skipping from membrane to membrane through the block. A replica is then made of the fracture face, usually by depositing carbon on it, and shadowing the replica with platinum, and the tissue is finally thawed and dissolved, leaving the replica to be viewed in the TEM. The IMPs usually remain embedded in the cytoplasmic half of the lipid bilayer, pulling out of the extracellular half. Figure 3.2 shows such particles in a freeze fracture preparation.

Many of the phospholipid molecules have no fixed position in the lipid bilayer, but are capable of sliding past each other: the whole membrane is capable of ballooning out or changing shape. In the same way, some protein molecules embedded in the membrane are capable of floating laterally in any direction, though they are unlikely to leave the membrane or rotate, turning somersaults. This freedom to move laterally in the lipid bilayer is an important feature of IMPs. It can obviously be limited if the ends of the proteins that project either from the extracellular or the cytoplasmic surface of the membrane are anchored to some structure lying parallel to the membrane.

The glycocalyx

The phospholipid bilayer with its IMPs is not the complete cell membrane: attached to the extracellular surface of the bilayer, extending out in branching patterns from it, are molecules which combine a protein or lipid base, that anchors them to the membrane, with complex chains of sugars (Fig. 3.1). This glycocalyx, as it is called, often appears on micrographs as a fuzzy coat outside the parallel dark lines of the membrane (Fig. 4.11). It appears to function in cell recognition, permitting or preventing contact between the membranes of adjacent cells. The extracellular position of this glycocalyx reminds us that the cell membrane has recognizable polarity; this can also be deduced from biochemical observations, such as those on ion transport by Na−K ATPase referred to above.

The cell membrane, then, defines the boundaries of the cell, creating and maintaining surprising differences in composition between the cell and the surrounding fluid. These include the creation of a difference in electrical potential between the inner and outer surfaces of the membrane, which is the basis of excitability in, for instance, nerve and muscle (to be discussed more fully in Chapter 12).

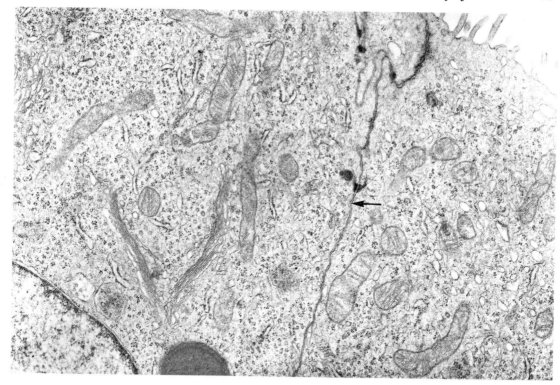

Fig. 3.3   Cytoplasm of two cells in the luminal epithelium of the rat uterus: cell membranes arrowed. Many membrane-bound structures are visible, including part of a nucleus (bottom left). (× 18 000.) (Courtesy, Dr J.B. Furness.)

Phospholipid membranes exist within the cell: wherever you see them, they define volumes of fluid which differ in composition from the fluid surrounding them (Fig. 3.3).

The nucleus

The nucleus is the most obvious membrane-bound structure within the cell. The nucleus contains the genetic material — in man, this is grouped into 46 chromosomes, each a complex of DNA with proteins. The activation of particular genes may result in considerable changes in cell behaviour — those for cell division are one example. We do not know the precise events that activate genes, but it makes good sense for the cell to control strictly the micro-environment inside the nucleus, so that activation becomes a "deliberate" act rather than the result of chance fluctuations in the composition of the fluid around the genetic material. The nuclear membrane, then, separates the genetic material from the hurly-burly of the cytoplasm.

The nuclear membrane

There is not just a single membrane around the nucleus, but a double one, enclosing a narrow space (Fig. 3.5). The separation of nucleus from cytoplasm is further emphasized

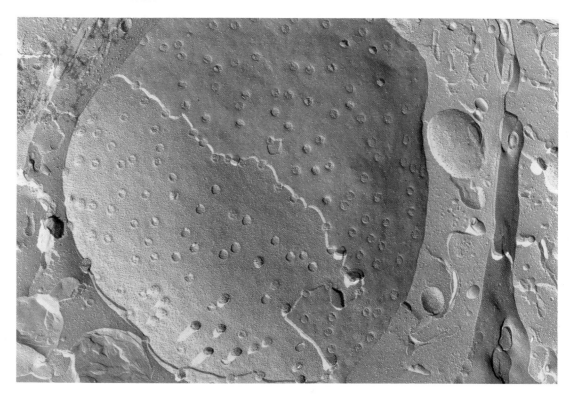

Fig. 3.4   *Freeze fracture preparation of nuclear membrane, showing many circular profiles of nuclear pores.* (× 25 500.) (Courtesy, Dr C.R. Murphy.)

by stacking condensed chromatin, representing genes not in use at the time, against the inside of the nuclear membrane. How, then, does the nucleus communicate with the cytoplasm? Freeze fracture shows many circular areas in nuclear membranes, called nuclear pores (Fig. 3.4), which, seen with the TEM, have quite elaborate organization (Unwin and Milligan, 1982). A ring or annulus of proteins surrounds a central, spherical particle. These pores are situated at gaps in the condensed chromatin, and close off the

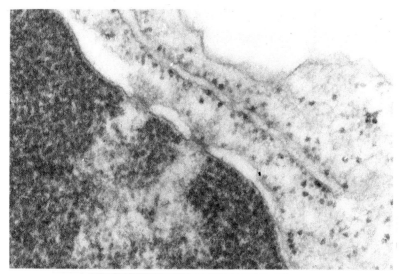

Fig. 3.5   *TEM of nucleus (bottom left), and nuclear membrane, showing perinuclear space and its closure at two nuclear pores. (× 76 500.) (Courtesy Dr M. Haynes.)*

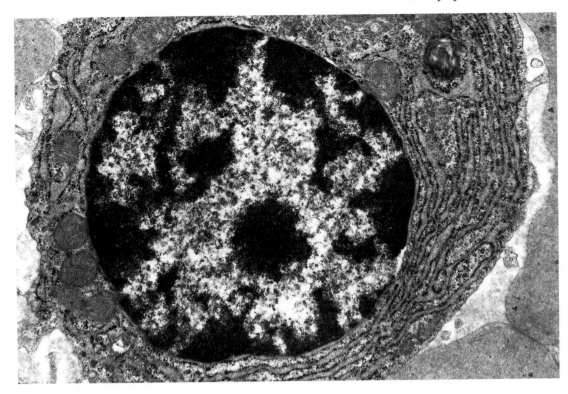

Fig. 3.6   TEM of a plasma cell, showing its nuclear structure. (× 13600.) (Courtesy, Dr M. Haynes.)

perinuclear space (Fig. 3.5). They thus bring the cytoplasm into close proximity to the interior of the nucleus.

The nuclear membrane provides a significant barrier to the free diffusion of ions and other small molecules. We are less certain how the nuclear pores function. Many signals from the cytoplasm that activate genes are in the form of proteins which migrate into the nucleus: the cytoplasmic receptors for steroid hormones are one example. Communication in the opposite direction is by means of RNA, synthesized in the nucleus and migrating out into the cytoplasm.

Chromatin

The nucleus itself contains chromatin, which may be inactive, in which case it is condensed and relatively deeply staining, or active, when it is dispersed in clear, watery regions that stain lightly. A densely stained structure, the nucleolus, is often seen somewhere in the centre of the nucleus. This contains filaments and granules in varying patterns, and is very rich in RNA. It is the site of production of new ribosomes.

Clearly, since inactive chromatin is condensed, a small, densely stained nucleus is not using many of its genes. By contrast, a large, pale nucleus suggests that much genetic material is being read. The presence of large or multiple nucleoli indicates the synthesis of many new ribosomes. Note the differences between nuclear and cellular activity. A plasma cell, synthesizing a specific antibody, will be metabolically very active, yet its nucleus looks "inactive" since the cell's functions depend on relatively few genes (Fig. 3.6).

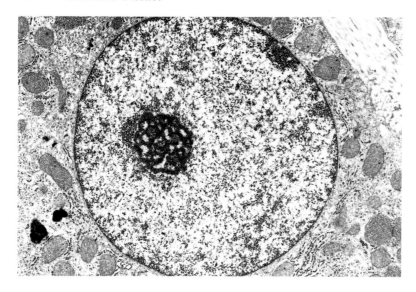

Fig. 3.7 TEM of the nucleus of a hepatocyte from human liver: the densely stained, coiled structure is the nucleolus. (× 5300.) (Courtesy, Dr M. Haynes.)

What can you predict about the cell whose nucleus is shown in Fig. 3.7? (Note 3.A)
How would this nucleus and the one of the plasma cell (Fig. 3.6) appear down the light microscope? (Note 3.B)

The cytoplasm

The cytoplasm contains many structures and organelles that can be recognized by electron microscopy (Fig. 3.3). Some of these, such as the mitochondria, the endoplasmic reticulum and lysosomes, are separated off from the rest of the cytoplasm by phospholipid membranes: we can at once assume that their contents differ considerably from that of the cytoplasm generally. Other structures, such as free ribosomes, glycogen granules and various filamentous proteins, lie within the cytoplasm itself. Our primary interest here is understanding how the cell functions, and how these functions are reflected in its structure, so, rather than listing the organelles of the cytoplasm, we will list its major functions, and discuss the appearances associated with them.

Energy production

Energy is provided within the cell in the form of the high energy phosphate bond of adenosine triphosphate (ATP). The details of the production of ATP from adenosine diphosphate (ADP) vary from tissue to tissue, and, within a particular tissue such as skeletal muscle, vary with physiological state. In general, about 20% of ATP production results from glycolysis in the cytoplasm, the remaining 80% arising from the activity of mitochondria. Within the latter, free fatty acids and the pyruvate produced by glycolysis are oxidized to carbon dioxide and water. The energy thus liberated is used to synthesize ATP from ADP — a process known as oxidative phosphorylation.

Mitochondria. Mitochondria have two distinct phospholipid membranes separating the tiny volume of water inside them from the cytoplasm (Fig. 3.8). The outer membrane is

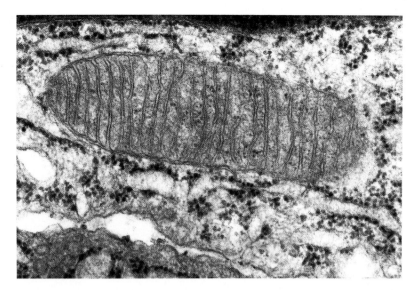

Fig. 3.8 TEM of a mitochondrion, surrounded by RER. (× 65 000.) (Courtesy, Dr M. Haynes.)

smooth and regular; the inner one follows its contours, but in addition is thrown into a number of deep folds, carrying it into the centre of the mitochondrion. These folds are called cristae. On the cristae, many subunits occur, like tiny mushrooms growing up into the central volume of the mitochondrial matrix: these subunits are believed to contain the enzymes responsible for oxidative phosphorylation. These subunits are very small, with heads about 10 nm in diameter.

How large would such structures be in Fig. 3.8? (Note 3.C)

Mitochondria are 2–6 μm long, and about 0·2 μm wide.

Can mitochondria be seen with the light microscope? (Note 3.D)

They seem in some ways to be almost independent structures, writhing and moving about in the living cell. They even increase in number by dividing, rather like bacteria, and contain filaments of DNA and their own ribosome-like particles of ribonucleoprotein. Cells with high energy needs are rich in mitochondria, which may not be randomly distributed in the cytoplasm. Membranes which are highly specialized for pumping ions against a gradient, for instance, require large amounts of ATP, and the foldings of cell membrane that provide large numbers of sites for ion pumping are associated with mitochondria lying between the folds (Fig. 3.9). Other sites within the cell with particularly high energy needs may also have high local concentrations of mitochondria.

Synthetic activity

There are two sorts of synthetic activity: in the first, cells grow and differentiate, building more cytoplasm and the organelles needed for specialized activity; in the second, cells synthesize a product which is secreted into the extracellular spaces or even on to the outer surface of the body. The markers which alert us to the first type of activity are the ribosomes required for protein synthesis. These lie free in the cytoplasm, often in small clusters, sometimes in considerable numbers. The only other clue we get to cell growth comes from examination of the nucleus.

What appearance would you expect to see in the nucleus of a cell that is growing and differentiating? (Note 3.E)

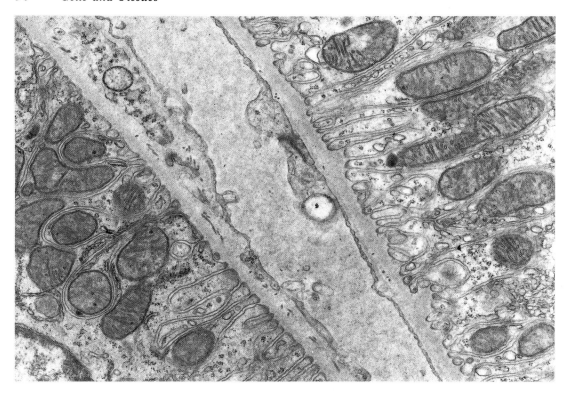

Fig. 3.9   TEM of a capillary (centre) with the basal portions of cells from the proximal convoluted tubules of the kidney on either side. Multiple infoldings of basal cell membrane are accompanied by mitochondria. (× 16 000.) (Courtesy, Dr D. Haynes.)

Endoplasmic reticulum. Material that is synthesized for export is not required within the cytoplasm of the cell in which it is made. If we think, for instance, of the digestive enzymes poured into the intestine, or of antibodies against infective agents secreted into the tissue spaces and circulating blood, they have no part to play in the economy of the cytoplasm of their cell of origin. The whole process of synthesis and preparation for secretion takes place in spaces separated from the cytoplasm by phospholipid membranes. These synthetic and secretory spaces vary in appearance, and carry several names as their functions change. The site of synthesis is called the endoplasmic reticulum (an alternative but less widely used name is ergastoplasm).

The endoplasmic reticulum (ER) is a complex of narrow, flattened passages (Figs 3.6, 3.10) which, rarely, is seen to communicate at one end with the perinuclear space. As material is synthesized, it is passed through the lining membrane into the ER. Ribosomes are, as we know, associated with protein synthesis, and, if the product being made has a protein component, the lining membranes are studded with ribosomes, regularly spaced (Fig. 3.10): this appearance is rough endoplasmic reticulum (RER). The synthesis of non-protein material for export, particularly steroids, is carried on in smooth endoplasmic reticulum (SER), which is also associated with detoxification of drugs.

The Golgi apparatus. The space inside the endoplasmic reticulum can be thought of as extracellular: once material is inside this space, it is out of the cytoplasm. But it seems

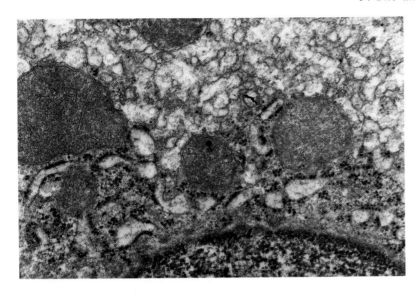

Fig. 3.10 TEM of cytoplasm from human hepatocyte, showing RER (centre) and SER (above). (× 53 500.) (Courtesy. Dr M. Haynes.)

likely that the initial products of synthesis are diluted by a considerable volume of water, and concentration occurs before secretion. Sugars may be added to some of the proteins, converting them to glycoproteins. These two activities, concentration and the addition of sugars to the synthesized material, take place in a specialized stack of curved, membrane-bound sacs called the Golgi apparatus (Fig. 3.11). This, while effectively an extension of the ER, is not in open continuity with it. Instead, small vesicles or droplets pinch off from the ER and travel to the outer, convex surface of the Golgi apparatus, fusing with it to

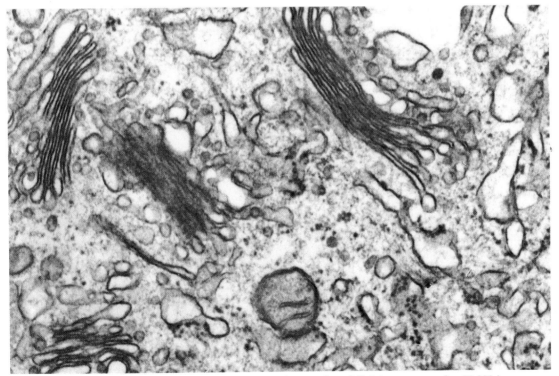

Fig. 3.11 TEM of cytoplasm, showing several examples of Golgi apparatus. (× 51 000.)

allow the contents to enter the Golgi saccules. This method of transport within vesicles is common within the cell, and provides a valvular mechanism, permitting material to flow from the ER to the Golgi, but not in the reverse direction. Each vesicle that leaves the ER for the Golgi apparatus reduces the amount of membrane available for synthesis, yet the total amount of ER in a synthetic cell does not diminish steadily. Clearly there is a balancing synthesis or recycling of membrane somewhere on the ER, perhaps at its communication with the perinuclear space.

Secretion granules and vesicles. As the synthetic product passes through the Golgi apparatus, then, it is concentrated and often modified. It leaves the inner, concave surface of the Golgi in a further series of small, membrane-bound transport vesicles, which have several possible fates. If secretion is continuous, these vesicles may pass directly to the cell membrane, fusing with it and inserting themselves into it as they open, setting free their contents outside the cell. But many cells, while synthesizing material all the time, only secrete intermittently, so that the product is liberated to coincide with other activities outside the cell. Plasma cells that make antibodies are examples of continuous secretion; the cells that make digestive enzymes in the pancreas are examples of intermittent secretion. In the latter case, the products of synthesis must be stored in the cell until the signal for secretion arrives. This storage takes place in secretion granules, which are membrane-bound (Fig. 3.12). The transport vesicles from the Golgi appear to coalesce to produce large, spherical droplets, in which the material for secretion becomes very densely packed, often in association with carrier proteins. These granules can become large enought to be seen with the light microscope, and are usually electron dense in the TEM. Their mechanism of secretion is, again, insertion into the cell membrane and opening to the exterior. Their presence within a cell is evidence that secretion takes place intermittently.

Inner and outer membranes. One further function of the Golgi apparatus has to do with the structure of the phospholipid membranes within the cell. So far, we have assumed that membranes within the cell have a common structure. However, evidence is accumulating that there are two classes of phospholipid membrane, inner and outer, the nuclear membrane an example of the former, the cell membrane an example of the latter. The nuclear membrane is continuous with the ER and the transport vesicles linking ER to the convex face of the Golgi apparatus: these are all lined with inner membrane. The concave face of the Golgi apparatus, the secretion vesicles and storage granules have membranes with similar structure to that of cell membrane. Transport vesicles of the inner type never fuse with structures that have outer membrane; secretion vesicles with outer membrane never fuse with and open into ER. Mitochondrial membrane appears to be of the inner type. The Golgi apparatus represents a watershed, accepting material that comes to it in inner membrane, and packaging it into outer membrane ready for fusion with the cell membrane (Farquhar and Palade, 1981).

The act of secretion inserts areas of membrane that once surrounded vesicles or granules into the cell membrane, yet the cell membrane does not increase in area with time in a secreting cell. There must be withdrawal of phospholipid in strictly regulated amounts from the cell membrane to preserve this balance, and recycling of the molecules through the Golgi apparatus, the source of the outer class of membranes.

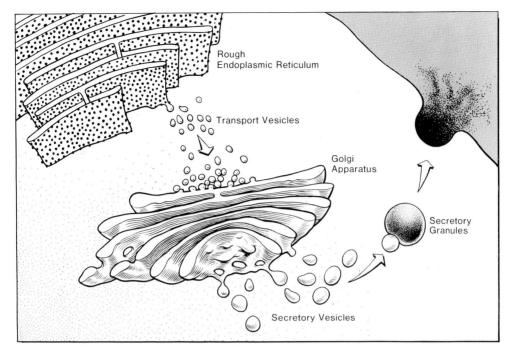

Fig. 3.12   Diagram of stages in the intermittent secretion of protein.

Absorption

Anything that enters the cell must pass through the cell membrane. Some molecules appear to diffuse through quite freely: steroid hormones, which are lipid soluble, come into this category. Others, particularly ions and sugars, can diffuse through channels in the phospholipid bilayer made by IMPs, as can water molecules. Some ions and small molecules are, in addition, assisted into or out of the cell by enzymes in the membrane, which form more of the IMPs seen on freeze fracture. So quite an extensive traffic takes place at the cell membrane in the absence of recognizable structural specializations. If this traffic needs to be greatly increased, as at the surface of cells facing the small intestine which absorb all the sugars and aminoacids taken in from the food, the cell membrane is thrown into many microvilli, effectively increasing the surface area greatly (Fig. 4.11). On such an absorbing membrane, the number of IMPs per unit area of membrane is high.

Phagocytosis and pinocytosis. Many substances are taken into the cell, even though they cannot readily pass through phospholipid membranes. This is achieved by two related mechanisms: pinocytosis (drinking by the cell) and phagocytosis (eating by the cell). In both cases, areas of cell membrane bulge into the cell and close off, forming separate vesicles lined with cell membrane. In pinocytosis, these vesicles are small and appear to contain nothing but extracellular fluid: they often transport proteins in solution across a cell, to liberate them at the opposite cell membrane. In phagocytosis, the vesicles are much larger, and may in addition to fluid contain recognizable material such as bacteria or fragments of cell debris. These two processes have only transferred the material from the cell surface to vesicles within the cytoplasm. The cell must somehow reduce the contents to

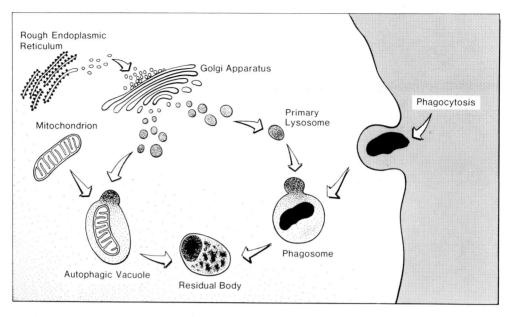

Fig. 3.13   The synthesis and fate of lysosomes. Autophagic vacuole, residual body and phagosome in this diagram are all secondary lysosomes.

a form that can enter the cytoplasm, in other words digest them, particularly in the case of phagocytosis.

Lysosomes. Digestive enzymes are proteins which must, for obvious reasons, be kept segregated from the cytoplasm in membrane-bound structures.
Where will these enzymes be synthesized? (Note 3.F)
Following synthesis, the enzymes will be concentrated and perhaps modified before being packaged in membrane-bound vesicles, in which they are retained in the cell.
What structures will be involved in this process? (Note 3.G)
Just as secretion granules fuse with the cell membrane, so vesicles containing digestive enzymes (lysosomes) fuse with the outer type of cell membrane that surrounds phagocytosed material, emptying their contents into the vacuole. The material in the vacuole is then broken down into fragments capable of crossing the membrane into the cytoplasm.

Any membrane-bound structure containing the digestive enzyme acid phosphatase is called a lysosome. By convention, this enzyme is used as a biochemical and cytochemical marker by which lysosomes are recogized. The clear vesicles produced at the concave face of the Golgi (Fig. 3.13) fit this description, but so do the large and varied phagocytic vacuoles after vesicles have fused with them and emptied their enzymes into them. The clear vesicles are called primary lysosomes, the phagocytic vacuoles secondary lysosomes; the latter group can be very varied in appearance, depending on the material originally taken in and how far digestion has progressed, so multivesiculate bodies and residual bodies are both names given to particular appearances of secondary lysosomes.

Some cells carry large primary lysosomes in preparation for phagocytosis of invading bacteria. As with secretory vesicles and secretion granules, these large primary lysosomes ready for intermittent use have a strongly staining, condensed appearance, by light and electron microscopy. The granules of the polymorphonuclear leucocytes of the blood (Fig. 7.3) are an example.

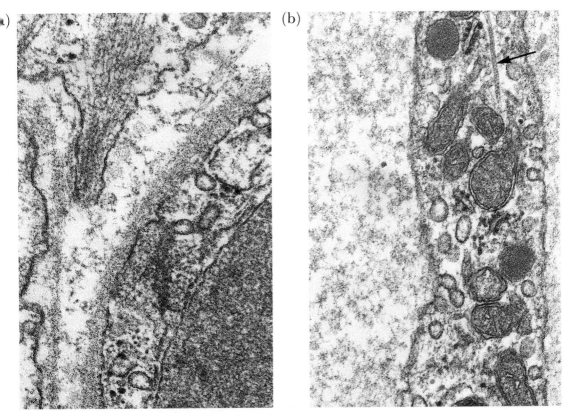

Fig. 3.14   Skeletal elements in the cytoplasm, seen by TEM. (a) Bundles of microfilaments in alveolar cell of possum lung (top centre). (× 66 500.) (b)   Microtubule (arrowed) in endothelial cell of alveolar capillary in possum lung. (× 46 000.) (Courtesy, Dr J. Fanning.)

Autophagosomes. Note that primary lysosomes have been packaged in the outer class of membrane by the Golgi, and can thus fuse with the membranes of vacuoles derived from cell membrane. Sometimes, the cell has to dispose of some of its own cytoplasmic contents, such as mitochondria or a volume of RER. But lysosomes cannot fuse with inner membrane. In a remarkable sequence of events, the cell assembles a new membrane of the outer variety around the unwanted structure, and primary lysosomes then fuse with this. Such a structure containing material derived from the cell itself is called an autophagosome or autophagic vacuole (Fig. 3.13), in contrast to the heterophagic vacuoles containing material from other cells, extracellular structures or invading foreign organisms.

The cell skeleton

We have dealt now with the main membrane-bound organelles in the cell: all that remains is the cytoplasm itself. The cytoplasm is an aqueous solution containing many proteins and smaller molecules, but even this portion of the cell is not a simple, structureless fluid. The TEM has revealed a number of filaments and tubules running through the cytoplasm, which appear to give it shape and a certain degree of rigidity, and are collectively known as the cytoskeleton.

Fine filaments, ranging from 7 to 10 nm in diameter, run in curving bundles through the

(a)   (b)

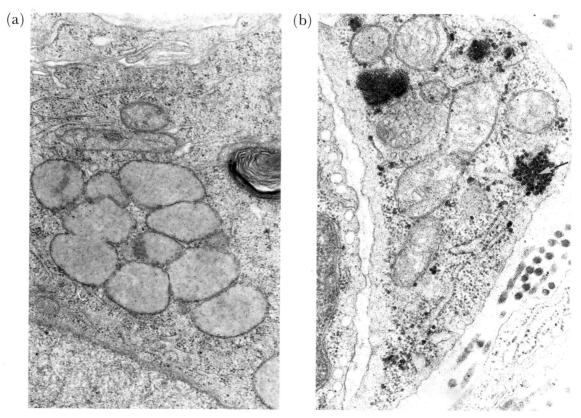

Fig. 3.15   Storage within the cytoplasm, seen by TEM. (a) Lipid droplets (× 17500). (b) Glycogen granules (arrowed), larger than ribosomes on RER, and irregularly clustered. (× 44000.) (Courtesy, Miss C. Lunam.)

cytoplasm (Fig. 3.14a), often seeming to converge on particular areas where a layer of dense material lies on the cytoplasmic side of the cell membrane. These filaments are protein, the majority of them actin. The part played by actin in contraction and cell movement will be considered in Chapter 9. In stationary cells, actin microfilaments help to maintain shape and provide resistance to deforming forces.

Microtubules are larger, with an outer diameter of about 25 nm and a wall 5–7 nm thick (Fig. 3.14b). They also are protein in nature, built up of repeating subunits, and they seem to add a certain stiffness to the cytoplasm. They also appear to be associated with the movement of organelles through the cytoplasm. The mitotic spindle is composed of microtubules. Microtubules appear to guide transport vesicles and secretory granules to their destinations. Some chemicals such as colchicine prevent the assembly of microtubules from their subunits; in addition to blocking mitosis by preventing the formation of the spindle fibres, these compounds disrupt the transport of vesicles within the cell.

Storage within the cell

Compounds may be stored in the cell in recognizable forms. Lipids occur as droplets of fat in the cytoplasm. Since these lipids are in a separate, non-aqueous phase already there is no point segregating them from the rest of the cytoplasm in a phospholipid membrane (Fig. 3.15a). Fats are a store of energy, and we shall see in Chapter 5 how some cells are given over entirely to this storage function.

Sugars may be stored as glycogen, appearing in the cytoplasm as particles of 20–30 nm diameter, often arranged in irregular clumps (Fig. 3.15b).

Some pigments in crystalline form may be seen in the cytoplasm. Lipofucsin is an irregular and heterogeneous mass, which is thought to be the indigestible remnants of lysosomal activity. If present in considerable amounts, it gives a tissue a yellow-brown appearance. It increases in amount in old age. Granular masses of haemosiderin may also occur in cells concerned in the breakdown of red blood cells: this is an iron-containing pigment derived from haemoglobin.

Linking cell structure to function

Our methods of observing cells inevitably influence our ideas about them, and nearly all the available techniques require the cells to be killed before we look at them. Cells are incredibly busy and active, with vesicles travelling in different directions, mitochondria writhing about more slowly, and everything shimmering with Brownian movement. Cells change shape, move, change direction, divide. Take every opportunity to see living cells, either on film or videotape, or directly in the light microscope.

What techniques make it possible to view live cells in the light microscope? (Note 3.H)

We can make many correlations between the structures recognized on sections and cell function. But many cells contain all the organelles and structures mentioned in this chapter. If we wish to identify the principal function of a cell, we must identify those structures that are present in usually high concentrations. The possession of a small length of RER does not mean that the main activity of a cell is the synthesis of protein for export: most cells have lysosomes, synthesized on the RER, and many, in addition, may release minute quantities of proteins into the extracellular fluid for signalling purposes. Identifying the principal function of a cell from its electron micrograph is a matter of judgement and experience, as well as the recognition of the various subcellular structures. Fortunately, cells carry no unnecessary baggage, and the fraction of the cytoplasm occupied by, for instance, RER or contractile proteins reflects accurately the cell's immediate need for these elements. This is the basis for morphometry, by which cell composition and structure can be measured to provide a quantitative basis for statments about its function.

Let us start acquiring experience by looking at two cell types, the fibroblast and the macrophage, relating their observed structure to their known functions. We shall go on then to look at a micrograph of an unknown cell type, attempting to predict its functions from its appearance.

The fibroblast

The fibroblast is responsible for the synthesis of many of the extracellular structures of the body: its main products include collagen, a protein, and glycosaminoglycans (GAGs), which are large molecules containing aminosugars, aminoacids and sugars. The functions of these components of connective tissue are discussed further in Chapter 5. The fibroblast

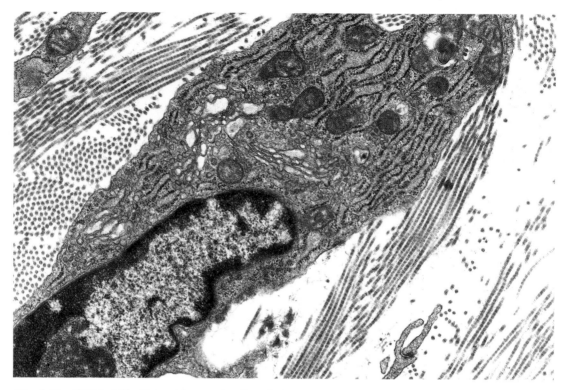

Fig. 3.16   TEM of fibroblast with many bundles of collagen fibres around it. (× 14 500.) (Courtesy, Dr J.B. Furness.)

(Fig. 3.16) is typically an elongated cell with a central, ovoid nucleus. The long axis of the cell appears related to the direction in which collagen fibres are oriented. It grows readily and is easily observed in tissue culture, away from the collagen that normally surrounds it. It can move slowly, and tends to form elaborate networks in the body, making contact with nearby fibroblasts by means of long processes.

The nucleus is moderately large, usually with an obvious nucleolus and not very much condensed chromatin.

The cell membrane is fairly smooth, without massive foldings, and there are several mitochondria and possibly a few lipid droplets, but neither in great abundance. The most significant feature of the cytoplasm is the number of lengths of RER, but there are also scattered groups of free ribosomes. One or two Golgi zones lie near the nucleus, and a number of vacuoles, some containing amorphous, rather fuzzy, material lie between the Golgi and the cell membrane. A few microfilaments and microtubules can sometimes be seen in the cytoplasm. There are no obvious secondary lysosomes, though some of the vesicles may be primary lysosomes.

This is the picture of protein synthesis, proceeding at a steady but not extraordinary rate, with the product being secreted continually — note the absence of secretory granules, which would have suggested intermittent secretion.

Such a cell would appear to lack very distinctive features at the light microscope level. There is not enough RNA in the cytoplasm to make it strongly basophilic, nor are there acidophilic secretion granules. In conditions of greatly increased collagen synthesis, such

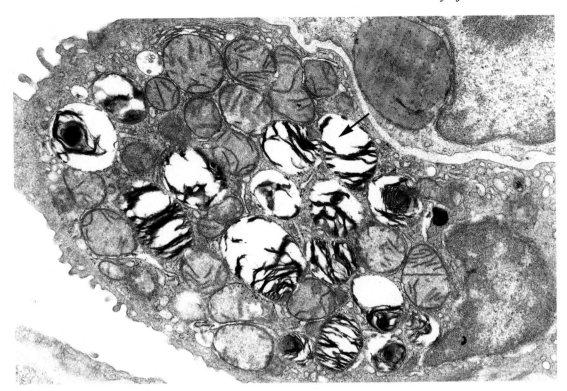

Fig. 3.17 TEM of a macrophage from human lung. The large, membrane-bound structures (arrowed) are secondary lysosomes. (× 14000.) (Courtesy, Dr D. Haynes.)

as wound healing, fibroblasts enlarge and the amount of RER increases considerably, producing cytoplasmic basophilia.

What is cytoplasmic basophilia, and what does it signify? (Note 3.I)

The collagen is produced and secreted into the extracellular fluid in precursor form, to be polymerized there into the characteristic fibres described in Chapter 5.

The macrophage

The macrophage (Fig. 3.17) is the cell responsible for the removal of debris from the extracellular space of the body. It is usually more rounded than the fibroblast, and capable of active movement. The nucleus is typically slightly smaller than that of the fibroblast, but rounded; it has one nucleolus and, again, not a great deal of condensed chromatin. The cytoplasm may have short processes, but nothing like those of the fibroblast. Its contents include many short lengths of RER and scattered free ribosomes, but the salient feature is the number of primary and secondary lysosomes and phagocytic vacuoles.

What is the difference between a phagocytic vacuole and a secondary lysosome? Are all secondary lysosomes alike? (Note 3.J)

Note that both cells contain the same list of structures — mitochondria, lipid droplets, tonofilaments, Golgi apparatus and so on. Yet they look quite different to the experienced eye. Relative frequences of RER and lysosomes and the overall shape of the cell are enough to distinguish them from each other.

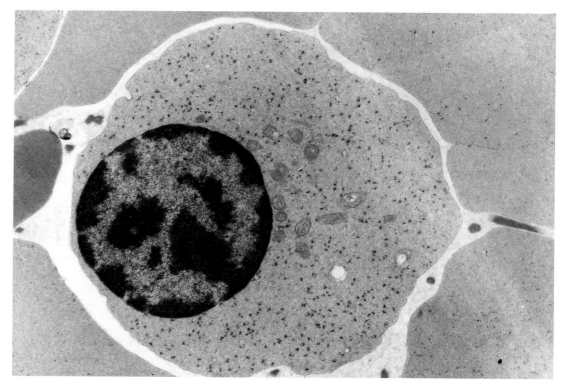

Fig. 3.18   TEM of a cell, for study. (× 11 000.) (Courtesy, Dr M. Haynes.)

The unknown cell

Figure 3.18 is an electron micrograph of a cell you have not seen before. Examine it, working through the nucleus, cell membrane and cytoplasm in order, and try to predict its major function. Finally, how would you expect it to look at the light microscope level? Answers are given in Note 3.K.

Further reading

Fawcett, D.W. (1981). "The Cell", 2nd Ed. W.B. Saunders, Philadelphia. An excellent atlas of cell ultrastructure, together with clear and up-to-date descriptions of each organelle and structure.

Hinckle, P.C. and McCartney, R.E.. (1978). How cells make ATP. *Scient. Am.* **238:3**, 104–123. The biochemistry, biophysics and ultrastructure of mitochondria.

Lodish, H.F. and Rothman, J.E. (1979). The assembly of cell membranes. *Scient. Am.* **240:1**, 38–53. How the polarity of the cell membranes is achieved.

Porter, K.R. and Tucker J.B. (1981). The ground substance of the living cell *Scient. Am.* **244:3**, 40–51. The cell skeleton of microfilaments and microtubules seen with the high voltage electron microscope.

Satir, B. (1975). The final steps in secretion. *Scient. Am.* **233:4**, 28–37. Clear illustration of the secretory process.

Sharon, N. (1977). Lectins. *Scient. Am.* **236:6**, 108–119. The structure of the glycocalyx and cell behaviour.

Weissmann, G. and Claiborne R. (Ed) (1975). "Cell Membranes." H.P. Publishing, New York. A multi-author book dealing not only with the structure of cell membranes, but their functions and biochemistry.

References

Farquhar, M.G. and Palade, G.E. (1981). The Golgi apparatus (complex) — (1954–1981) — from artifact to center stage. *J. Cell Biol.* **91**, 77s–103s.

Unwin, P.N.T. and Milligan, R.A. (1982). A large particle associated with the perimeter of the nuclear pore complex. *J. Cell Biol.* **93**, 63–75.

4

Epithelia: the body's limits

The boundaries of our bodies, the limits that mark us off from the outside world, are formed by a continuous sheet of cells, which separate the constant and controlled internal environment of the body from the varying and often uncontrolled external one. This sheet of cells has an extraordinarily complex shape. On the surface of the skin it dips down into sweat glands and hair follicles; at the lips, it is continuous with the lining of the mouth and on into the gut and the lungs, while in each of these in turn smaller or larger glands open to the surface (Fig. 4.1). This sheet of cells lines many spaces deep within the body, small pockets of the outside world which we for our own convenience have surrounded. Wherever such spaces are ultimately continuous with the outside world, in the urinary bladder, the uterus, the air sinuses of the skull and so on, they are lined by part of this continuous sheet of cells which is called epithelium.

Clearly the epithelium which lines the bladder faces a very different environment from that of the skin, and has different functions; its structure is also different, reflecting functional needs. So the continuous epithelial sheet varies from place to place in its organization, sometimes changing quite abruptly as it passes from one external environment to another. Underlying all these variations in structures, however, are certain features common to all epithelia.

Definitions

Definitions are difficult to apply strictly to the incredible variety shown by cell populations. Whereas everyone would agree that the continuous sheet of cells described above should be called epithelial, many apply the term to cell populations that are not part of that sheet. In the thyroid gland, for example, are follicles lined by cells which develop from the epithelium of the floor of the mouth during embryonic life. These follicles, which do not communicate with the exterior, can be regarded as separate droplets of the outside world, and the cells lining them may be called an epithelium. In the anterior pituitary gland or adenohypophysis, there is another cell population developed from the epithelium of the mouth of the embryo. This group of cells has also lost all contact with the surface of the body, and, in humans, does not surround a cavity. Are these cells epithelial? The whole of the central nervous system is developed from the surface layer of cells making up the ectoderm in the very early embryo. If we call the cells of the anterior pituitary epithelial, should those of the brain and spinal cord also be classified in this way?

42

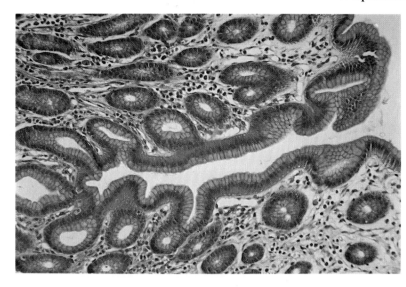

Fig. 4.1   Glands in the wall of the pyloric region of the stomach from a monkey. H & E. (× 160.)

Some histologists include as epithelia the cell layers lining the insides of blood-vessels and of the body cavities, such as the peritoneum and the pleura. These layers separate strictly regulated compartments within the body rather than facing an uncontrolled and potentially hostile external environment. The terms endothelium and mesothelium respectively are more appropriate.

In this chapter, we shall define epithelia as the continuous layer of cells separating the body from the external environment. This definition includes glands with cavities that communicate with the exterior by ducts; it does not include cells of epithelial origin that have lost contact with the exterior.

Features common to all epithelia

By virtue of their position at the surface of the body, facing the external environment, all epithelia share a number of features, which help us not only to understand their function, but to recognize them down the microscope.

Cellularity. Epithelia are composed entirely of cells (Figs 4.1, 4.2). In the tissues beneath, cells are often widely separated, with extracellular fluid, fibres, even the matrix of bone or cartilage lying between them. This is not true of epithelia. We face the outside world with a continuous sheet of cells so that nothing can enter or leave the body without passing through the monitoring systems of a living cell. There are no extracellular fibres in epithelia, no blood-vessels; the extracellular space itself is as small as possible and actually eliminated in places between adjacent cells.

Polarity. Epithelia show polarity. The outer surface faces the external environment, the inner suface a controlled, internal one, and the cells of the epithelium reflect this difference. Cells that are not epithelial often show no difference in structure on one or other surface. Epithelia are always polarized along the axis from external to internal environments.

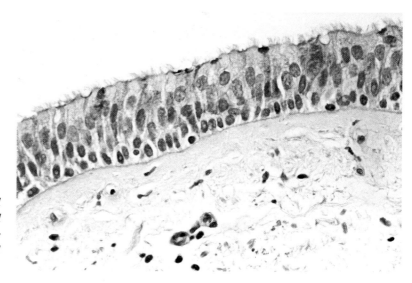

Fig. 4.2 Human bronchial epithelium (above) and underlying connective tissue (below), separated by a basal lamina. PAS stain. (× 420.)

Specialized contacts. Epithelial cells have specialized contacts with their neighbours to hold the cells together and reduce or eliminate the extracellular space. In an epithelium one cell thick, specializations for attachment link adjacent cells at their lateral membranes near to the luminal surface (the lumen is the cavity of any hollow organ or tube). In this position they form the junctional complex (Fig. 4.3), comprising the tight junction (zonula occludens) nearest the lumen, the intermediate junction (zonula adherens) and the desmosome (macula adherens) below. These are described in more detail below. Apart from these specialized structures, adjacent walls of epithelial cells often fit together in interlocking patterns, like a jigsaw puzzle, which may well be a further device for holding fast.

Lateral communication. Epithelial cells communicate freely with their neighbours by specialized contact areas on their membranes, allowing the spread of information laterally along the cell layer. These communication pathways are known as gap junctions or low resistance junctions. They are modifications of the lateral cell membranes which allow water, some ions and many small molecules to pass from the cytoplasm of one cell to that of its neighbour. The structure and functions of these will be discussed later.

Basal lamina. Epithelia rest on and are attached to a fine carpet of tough fibres, called a basal lamina, on that surface which faces the internal environment of the body. The basal lamina appears to be made by the epithelial cells themselves. It consists of very fine fibres of collagen, often called reticular fibres, set in a matrix of glycosaminoglycans: both of these will be discussed in detail in Chapter 5. The basal lamina (Fig. 4.2) seems to have several functions. It provides physical support to the epithelium resisting stretching and tearing forces; the arrangement of the connective tissue beneath the basal lamina often strongly reinforces this support. It also limits contact between epithelial cells and other cell types in the tissue spaces beneath. Only an occasional nerve fibre or a wandering white blood cell crosses this boundary into the epithelium. The basal lamina acts as a filter, allowing water and small molecules to pass through but holding back larger ones such as proteins. This certainly happens in the glomeruli of the kidney where the fluid that will become the urine is produced as a filtrate of blood plasma, and it may well happen in other

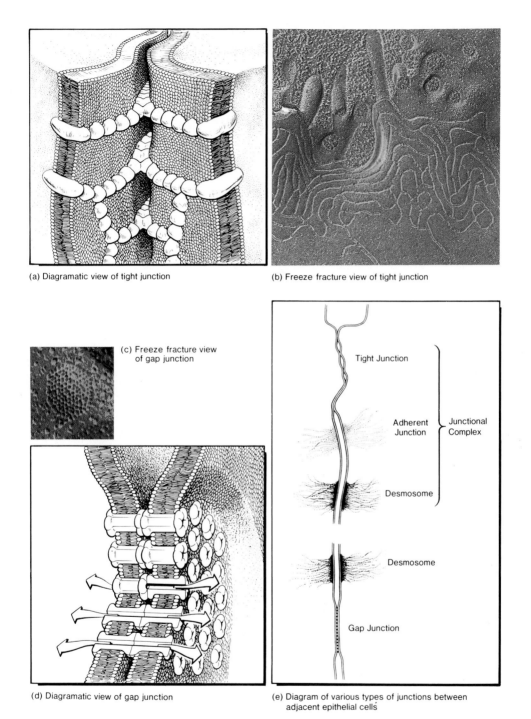

(a) Diagramatic view of tight junction

(b) Freeze fracture view of tight junction

(c) Freeze fracture view of gap junction

(d) Diagramatic view of gap junction

(e) Diagram of various types of junctions between adjacent epithelial cells

Fig. 4.3   Various types of contact between adjacent epithelial cells. (a) and (d) show the phospholipid bilayers of adjacent cell membranes. (b) and (c) are from cells lining the lumen of the rat uterus. (b, × 75 000: c, × 162 500.) (Freeze fracture pictures by courtesy, Dr C.R. Murphy.)

epithelia, restricting the range of molecules that can pass between the cells beneath and the epithelium. The basal lamina defines the space that may be occupied by epithelial cells; one important diagnostic feature of cancer cells derived from epithelia is their lack of respect for this boundary, which they breach to invade the tissues beneath.

Cell replacement. Finally, all epithelia face an outer environment which is to a variable extent uncontrolled and hostile. There may be bacteria or pollutants present, such as cigarette smoke, acid, concentrated urine or digestive enzymes. The more hostile the environment, the higher the probability of cell death and the greater the need for cell division to replace the losses. But there are immediate problems with cell replacement in a hostile environment. In cell division, the genetic code must be safely and accurately copied and partitioned between the daughter cells, a process that requires the closing down of genetic control mechanisms for several hours and the disappearance of the nuclear membrane. Dividing cells are thus very vulnerable, while highly specialized cells with cytoplasmic defences against the external environment are often committed to programmes that exclude cell division. Many epithelia are organized to shield dividing cells from the exterior. Those that permit cells in contact with the exterior to divide face a relatively mild and controlled environment.

The recognition of epithelia

These features provide us with a basis for recognizing epithelia in tissue sections. While it may not always be easy to identify the lumen, which may be reduced to a tiny space in some glands (Fig. 4.12), the cells themselves are tightly packed, with no significant extracellular space, no blood-vessels, no fibres or other non-cellular material in the epithelium itself. The basal lamina separates these cells from the cells, spaces and fibres of connective tissues. The epithelium is polarized, changing in structure from basal lamina to lumen, even when only one cell thick. With the electron microscope, junctional complexes and gap junctions may be seen in transmission or freeze fracture pictures; their presence can often by inferred at the light microscope level from the very close packing of the cells and their striking similarity to their neighbours, which may reflect the flow of information freom cell to cell. Mitotic figures may be seen (Fig. 4.10), and their frequency and position suggest how hostile the outside environment is at that point.

The organization of glands

We are familiar with epithelia as a sheet of cells separating the interior of the body from the outside. Figure 4.1 illustrates how this sheet is often folded into pits or glands, lying beneath the surface of the main epithelial sheet, but each with a lumen that communicates with the exterior. Such glands are called exocrine, and are a device for delivering on to the surface of the main epithelium a greater volume of secretory material or a different type of secretion than the cells of the surface epithelium can produce.

If we visualize a flat epithelial surface, the simplest possible gland is a single cell or cluster of cells in the epithelium itself, secreting on to the surface (Fig. 4.4). If more secretion is needed at that point on the epithelium than these cells can conceivably produce, the next level of complexity is a shape like a test-tube, lined with secreting cells, and opening on to the surface. More secretion still can be achieved by increasing the number of

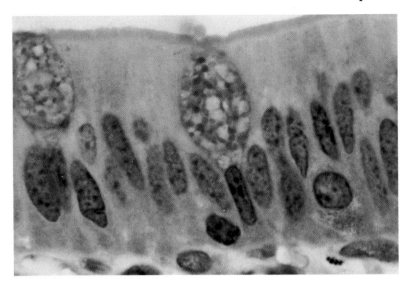

Fig. 4.4   *A goblet cell (centre) in the epithelium covering a villus, from the jejunum of a monkey. H & E. (× 1350.)*

secreting cells lining the gland, but, since it is seldom convenient to accommodate very long, straight glands, such an increase in cell number may produce several patterns. The duct may become elaborately coiled. Alternatively, it may branch, with a number of test-tubes lined with cells opening into the one duct, as in Fig. 4.1. This branching may continue through many generations of divisions in the duct system, each terminal duct opening into a small space surrounded by secreting cells. In the pancreas, for example, or the salivary glands, impressive volumes of fluid can be delivered to the surface of the duodenum or the mouth by glands which, in the case of the pancreas, may be 25 cm long and weigh over 100 g.

A single, unbranched gland is called simple, and it may be straight or coiled (Fig. 4.5). If it is the same diameter all the way down, it is tubular; often, however, the deepest part of the gland is expanded to form a spherical knot of cells around a central cavity, and this pattern is called acinar or alveolar, the terminal part being an acinus or alveolus. The larger glands have clearly differentiated ducts connecting the secretory portions to the surface. If such a duct branches, the gland is called compound. Once again, it may be tubular, either straight or coiled, or it may be acinar. Note that all the cells in duct and secretory regions are epithelial, and that the lumen communicates directly with an exterior surface.

Some groups of cells have no contact with the outer surfaces of the body, and produce secretions which travel directly into the surrounding blood-vessels. Such glands are called endocrine, their products are hormones. The major characteristics of such glands include the absence of any duct system, and an unusually rich blood supply.

Three further terms that you may meet in connection with glands describe the mechanism of secretion of the individual cells. Merocrine secretion involves the liberation of the synthesized material on the surface of the cell, which remains intact. This is usually achieved by the fusion of a membrane-bound vesicle or granule with the cell membrane. In some cells, notably those of the lactating mammary gland, the apical cytoplasm, which includes secretory products, is pinched off into the lumen: this is called apocrine secretion. Finally, the whole cell may break down and be secreted, as in sebaceous glands of the skin: this is holocrine secretion.

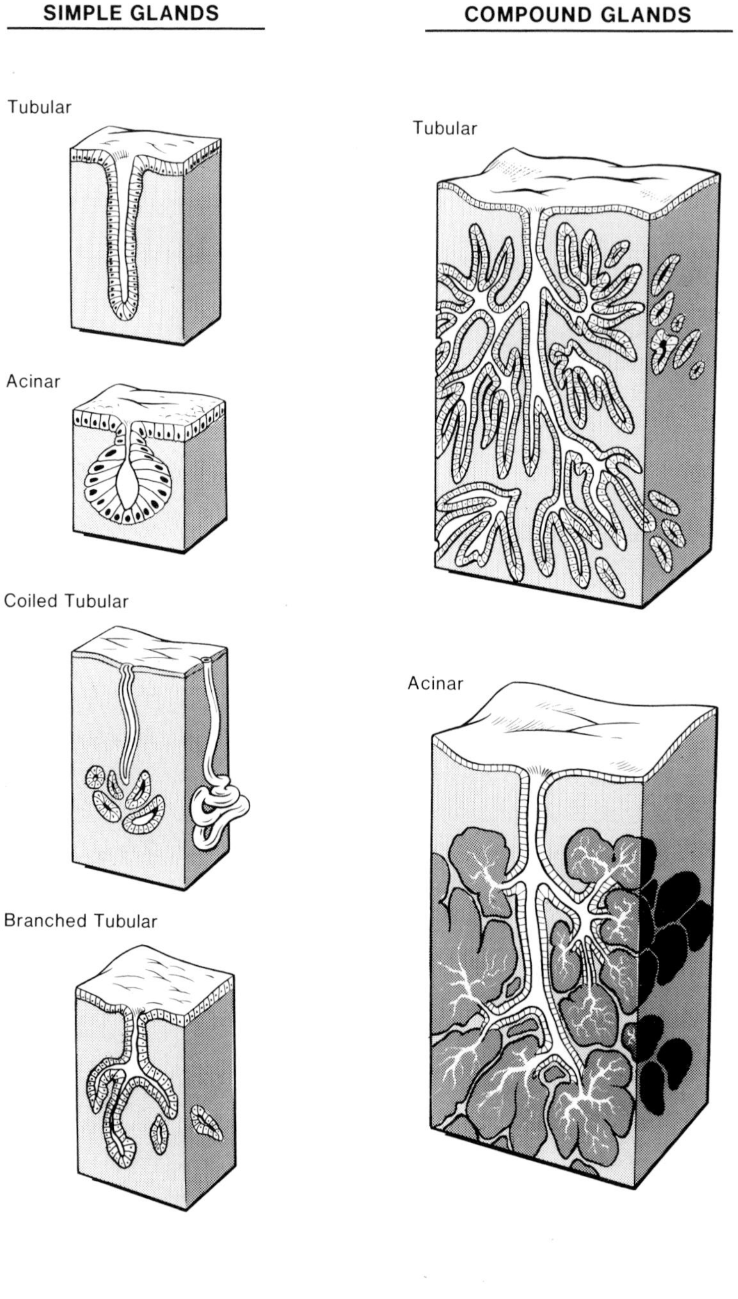

SIMPLE GLANDS

Tubular

Acinar

Coiled Tubular

Branched Tubular

COMPOUND GLANDS

Tubular

Acinar

Fig. 4.5   The classification of glands.

Contacts between epithelial cells

If a complete layer of cells is to be maintained between the external environment and our internal tissues in all conditions of movement and activity, contacts between epithelial cells must be capable of holding the membranes tightly and permanently together. This is the function of the junctional complex and the desmosomes. A further characteristic of epithelia, lateral communication, is carried out by gap junctions. Many other cell populations in the body have one or other type of specialized contact: few have the full complement seen in epithelia. These contacts will be described as they occur in epithelia which are one cell thick.

In many epithelia, the junctional complex can be seen with the light microscope as a thickening of the line marking the boundaries of adjacent cells close to the lumen: the old histologists called this appearance *terminal bars*. It may even be possible in some preparations to recognize the dense feltwork of microfilaments at the cell apex as a *terminal web*.

The junctional complex

Around the lateral surface of epithelial cells, near to the lumen, lies the junctional complex (Fig. 4.3). This can be resolved into three components: the tight junction or zonula occludens, the intermediate junction or zonula adherens and the desmosome or macula adherens. Tight and intermediate junctions rarely occur between neighbouring cells except in epithelia. Desmosomes, on the other hand, not only occur at many points between the lateral membranes of epithelial cells, but also in many other cell populations throughout the body.

The tight junction. The tight junction is a continuous band running round the neck of the cell and linking it to every adjacent cell: it lies nearer to the lumen than the other components of the junctional complex. Seen by the TEM (Fig. 4.3e) it is a zone where the outer leaflets of the membranes of adjacent cells appear to fuse together, with complete elimination of the extracellular space. A better picture of the tight junction comes from freeze fracture studies (Fig. 4.3b).
Why should this be so? What is the likely plane of fracture, and what structures does the technique reveal? (Note 4.A)
By this technique, tight junctions appear as branching and converging lines raised above the level of the fracture plane, rather like low walls around irregular fields seen from the air. These ridges are continuous lines of IMPs which link firmly to corresponding lines of IMPs in the neighbouring membrane, holding the two membranes tightly together and eliminating the extracellular space. Small marker molecules introduced into this space, whether from the luminal or basal side, cannot pass these ridges. The outside environment is thus sealed off from the interior of the body. Any molecule entering or leaving the body can only do so through the monitoring systems of a living cell.

The tight junction divides the cell membrane into two quite distinct regions, separating the membrane that faces the outside world from that facing other epithelial cells and the basal lamina. Evidence is accumulating that the structure and functions of the membrane may also differ in these two regions. The density of IMPs, the presence of microvilli and cilia, the abundance of the glycocalyx and the presence and direction of many transport

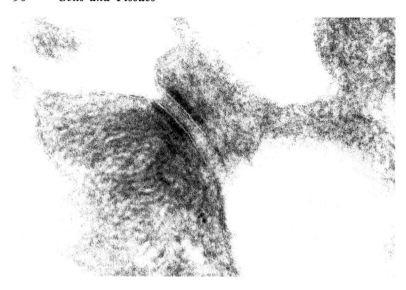

Fig. 4.6   TEM of a desmosome joining processes of two cells from the connective tissue of the rat uterus. (× 138 000.)

systems in the apical membrane differentiate it from membranes elsewhere on the cell (Cereijido *et al.*, 1978).

The intermediate junction. This is also a continuous band around the neck of the epithelial cell: it lies immediately below the tight junction. In TEM pictures, the membranes of adjacent cells lie parallel to each other, separated by about 20 nm. There is a dense thickening on the cytoplasmic face of the membrane, from which many fine microfilaments run out into the apical cytoplasm. No specialized structure can be seen between the two membranes, and freeze fracture studies fail to demonstrate characteristic densities or patterns of IMPs. The intermediate junction is believed to be a zone of firm adhesion between neighbouring cells, though we know very little of the forces that actually hold the cells together at this position. The closeness of tight and intermediate junctions suggests that they somehow support each other. Forces tending to peel two adjacent cells apart, starting at their base, would, if unchecked, risk pulling the tight junctions apart or tearing the cell membranes. The intermediate junction resists this peeling apart immediately next to the tight junction. The microfilaments anchored to the intermediate junction form a dense web at the apex of many epithelial cells, reinforcing and strengthening the cell at this point.

Desmosomes. Unlike the other two components of the junctional complex, desmosomes are discrete, circular spots on the membranes of adjacent cells. They occur in numbers just on the basal side of the intermediate junction, but also throughout the lateral membranes of epithelial and non-epithelial cells (Fig. 4.6). They are like spot welds, holding the membranes together. In TEM pictures, they resemble the intermediate junctions in many respects. Adjacent cell membranes run parallel and 20 nm apart, and appear very thick and dark because of a dense layer of material on the cytoplasmic side of the membrane. Fine microfilaments are attached to the membrane here, branching out into the cytoplasm and anchoring the desmosome to the structural skeleton of the cell. Unlike the intermediate junction, the desmosome is often seen to contain a line of dense material in the extracellular space, parallel to the membranes, and there are hints of fine fibrils crossing

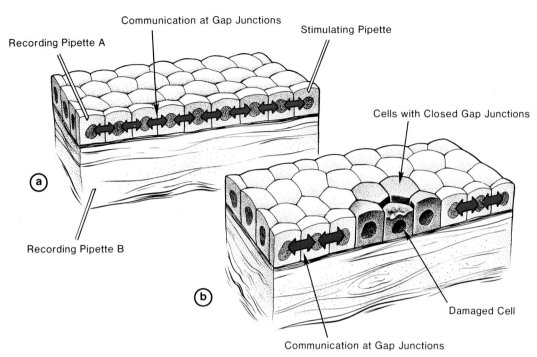

Fig. 4.7   (a) Experiment designed to demonstrate presence of low resistance junctions in an epithelium: for details see text. (b) Closure of low resistance junctions around a damaged cell.

the space from one membrane to the other. No specialized structure is seen on freeze fracture, however. Once again, the precise mechanism of attachment at desmosomes is not fully known.

The gap junction

If micro-electrodes are placed in two cells some distance apart in a continuous epithelial sheet, electrical pulses injected into the one can be detected in the other: the cells are electrically coupled (Fig. 4.7). No pulse can be detected in the luminal fluid or beneath the basal lamina. If small dye molecules, such as fluorescin, are injected into one epithelial cell, they will spread to adjacent cells but not into the extracellular space. This rapid flow of information suggests that the lateral cell membrane is permeable to some ions and small molecules, which cannot, however, enter the extracellular space. Special contacts, called *low resistance junctions*, must somehow couple the cytoplasmic compartments of adjacent cells.

There is now quite good evidence that gap junctions are the sites of this lateral communication. Gap junctions are not limited to epithelia. They are separate, discrete regions of membrane: in TEM pictures, adjacent membranes run parallel and very close to each other, separated by only 2 nm. Freeze fracture shows a closely packed array of IMPs in a honeycomb-like pattern. These are thought to be large protein molecules extending through the membranes of both cells, forming channels through membranes and ex-

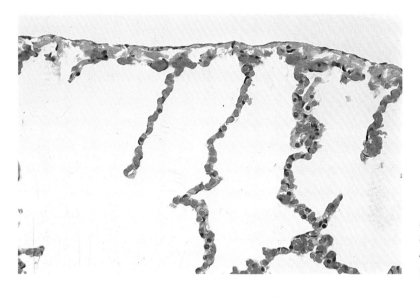

Fig. 4.8 Alveoli of monkey lung: the parietal pleura is at the top of picture. H & E. (× 250.)

tracellular space, linking the two cytoplasmic compartments. These channels are like an array of small pipes linking two tanks of water: dissolved materials can diffuse from one tank to the other without ever entering the air space between them (Fig. 4.3d).

The existence of this system of information flow between epithelial cells probably allows them to respond in a co-ordinated fashion to events happening at one point. It is also potentially dangerous, providing a channel for toxic substances to spread across the epithelium. Interestingly, a protective mechanism has evolved to close these channels if cell damage occurs. Ionic calcium is present at only extremely low concentrations in the cytoplasm of most cells. Damage to the cell membrane causes a rise in the concentration of calcium, which flows in from outside the membrane. Such a rise immediately closes the gap junctions of the damaged cell, isolating it from its neighbours (Fig. 4.7). Its neighbours also close their junctions with surrounding cells, forming around the damaged one a ring of undamaged cells, which are not in communication with the rest of the epithelial sheet. Restoration of a suitable concentration of ionic calcium restores communication (Loewenstein and Rose, 1978)

The classification of epithelia

The epithelial sheet covering the surfaces of the body varies in thickness: at one extreme is the layer which lines the alveoli or airsacs of the lung, too thin to distinguish with the light microscope (Fig. 4.8); at the other is skin several millimetres thick on the sole of the foot. In general, surfaces at which materials are transported into or out of the body have a relatively thin epithelium; those designed to prevent the passage of materials are thick. So the most basic classification of epithelia rests on the number of cell layers present. A *simple epithelium* is only one cell thick; a *compound epithelium* is composed of more than one layer of cells (Fig. 4.9).

Next, the shape of the cells in the outermost layer is used as a basis for classification. These may be large in surface area, but flattened, thin. They are called *squamous* (*squama* in Latin is a scale, as on a fish or snake). Cells may be approximately square, or *cuboidal*.

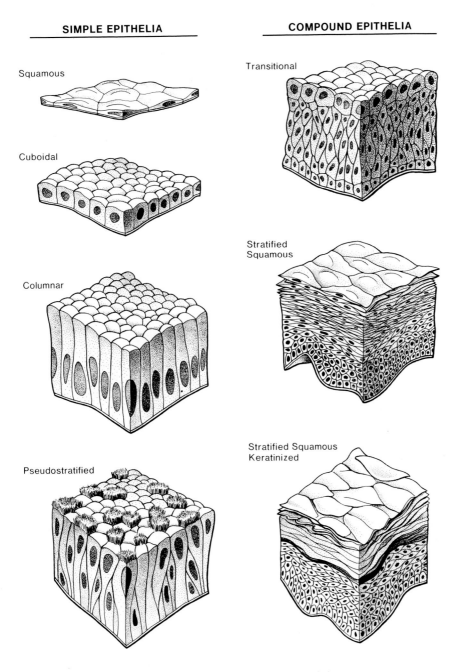

SIMPLE EPITHELIA

Squamous

Cuboidal

Columnar

Pseudostratified

COMPOUND EPITHELIA

Transitional

Stratified
Squamous

Stratified Squamous
Keratinized

Fig. 4.9 The classification of epithelia.

(a)

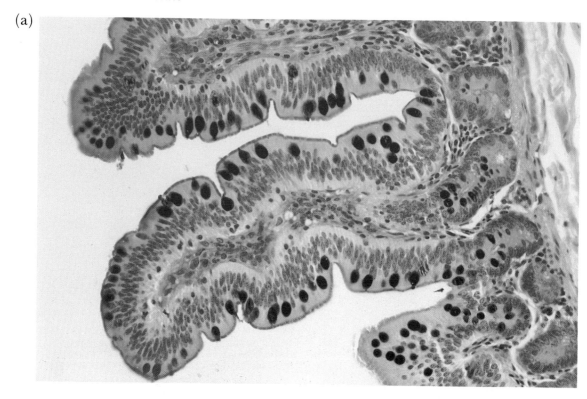

(b)

(c)

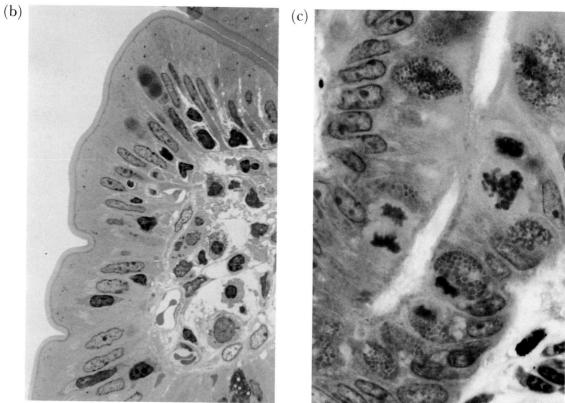

Fig. 4.10   The epithelium lining the small intestine. (a) General view with villi protruding into the lumen (left): goblet cells and surface mucus stained magenta. PAS. (× 345.) (b) Epithelium over the villus. H & E. (× 960.) (c) Epithelium in the crypts, with three mitotic figures. H & E. (× 1120).

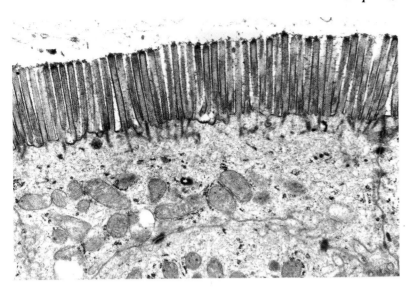

Fig. 4.11   TEM of micro-villi at the luminal surface of an absorptive cell on the surface of an intestinal villus. (× 20 000.)

Yet others may be much taller from basal lamina to lumen than their width, in which case they are *columnar*. A simple squamous epithelium consists of a single layer of flattened cells, like the lining of the alveoli of the lung. A stratified squamous epithelium has many layers of cells, the outermost of them flattened.

This two-stage classification covers all the epithelia of the body, with the addition of two further terms. Some epithelia have nuclei at two or more levels, when viewed with the light microscope, and so look stratified. At high magnification, in particular, by electron microscopy, every cell present rests on the basal lamina, but only some of them extend all the way to the lumen. This pattern, made up of columnar cells with smaller, rounded cells at their base, is called *pseudostratified*. It is a subgroup of the simple epithelia.

The second term which requires explanation is *transitional*. This is applied to a type of epithelium seen in the urinary tract: it is stratified, with surface cells large in area like squames, but thick and irregular. This highly specialized epithelium has evolved to face hypertonic urine. The term, transitional, is not descriptively useful and some histologists prefer to call this an urothelium.

A look at particular epithelia

Let us look at the epithelial sheet covering the body at six sites, to illustrate the general principles already stated and to see how the functional needs of a surface dictate the structure of the epithelium present.

The small intestine

More accurately, Fig. 4.10 shows a section of the jejunum, the second part of the small intestine where digestion of food is completed and much of the work of absorption carried out. It is not difficult to identify the epithelium. The layer next to the lumen is composed of *cells tightly packed*, shoulder to shoulder without extracellular fibres or spaces. At high magnification, small spaces exist between the bases of the cells, as in many epithelia where absorption is important, but the apical parts of the cells are tightly packed. It is a simple

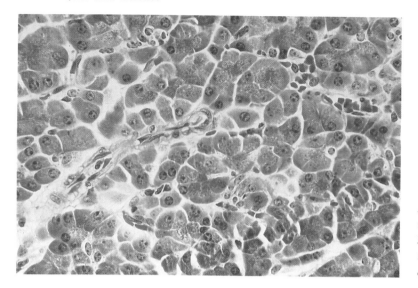

Fig. 4.12 Pancreas, show-
ing sections through many
secretory acini, set in connec-
tive tissue. Masson's tri-
chrome. (× 300.)

columnar epithelium, and the cells are clearly *polarized*. Even with the light microscope the cell surface facing the lumen has a fine striated appearance, the brush border, which the electron microscope resolves into a tightly packed mass of microvilli (Fig. 4.11), while the nuclei are nearer the base than the lumen. *Junctional complexes* (Fig. 4.3e) can just be distinguished as terminal bars with the light microscope (Fig. 4.10b). The curiously uniform appearance of the cells suggest the presence of *gap junctions* and shared informa- tion. The *basal lamina* separates epithelium from underlying cells and tissue spaces.

This sheet of cells does not just line the inside of a simple tube, but it covers large numbers of finger-like projections, or villi, which stick out into the lumen. The effect is to increase greatly the surface area available for absorption. The microvilli on each absorptive cell have a similar function, producing a vast area of surface membrane across which food materials can be transported. The presence of microvilli on a membrane always suggests absorption. Around each villus, numbers of small, test-tube shaped glands run down into the wall of the intestine, each with its lining of simple columnar epithelium.

The external environment faced by the cells on the villi is about as hostile as can readily be imagined. Within the intestine, cells from plants and animals are being rapidly converted to sugars, aminoacids and other small molecules by a spectrum of digestive enzymes, quite capable of doing the same for the epithelial cells themselves. Yet the epithelium has to be simple to permit absorption: a thick, protective epithelium would be useless here. The body has evolved several mechanisms to preserve intact this delicate layer of cells. First, a thin, continuous layer of mucus covers the epithelium (Fig. 4.10a). Mucus slows the diffusion rate, particularly of large molecules such as digestive enzymes, and helps to protect the cells from rapid attack. The cells that produce the mucus can be seen scattered throughout the epithelium and often distended with a droplet of mucus in the apical cytoplasm: they are called goblet cells.

In spite of this protective layer, the epithelial cells over the villi have a high probability of being damaged by the luminal contents. The whole sheet of cells is therefore replaced regularly, once every 36 hours in the mouse, every 3–5 days in humans. The dividing cells are kept segregated in the glands, or crypts, away from the hostile environment outside. Many mitotic figures can be seen in the crypts (Fig. 4.10c): in fact, this is the most rapidly dividing population of cells in the body. Cell production here causes cells to migrate up out of the crypts on to the bases and sides of the villi, pushing in front of them existing epithelial cells. These are finally shed from the tips of the villi into the lumen where they are digested and their raw materials reabsorbed by their successors. The whole epithelium is moving slowly towards the tips of the villi, new cells produced in the sheltered crypts replacing the old, which are shed and lost.

The pancreas

The pancreas is a solid-looking mass of tissue lying next to the first part of the small intestine, the duodenum. At first glance, it has little to do with an exterior surface, yet it communicates by a duct with the lumen of the intestine. Followed back into the pancreas, this duct branches repeatedly, each branch ending ultimately in a small space surrounded by epithelial cells. These cells manufacture digestive enzymes which pass in inactive form through the duct system to the duodenum. Here, the enzymes become activated by substances produced by the epithelium of the duodenum. The secretion of pancreatic enzymes is intermittent, so that their arrival coincides with the presence of food in the duodenum. The design of the pancreas provides a vast surface area for the synthesis and secretion of these enzymes into the gut.

Study the photomicrograph of pancreatic epithelium (Fig. 4.12).
How large would a red blood cell be at this magnification? (Note 4.B)
Given that digestive enzymes are proteins and that secretion is intermittent, what can you predict about the appearance of these cells in the electron microscope? What features at the light microscope level confirm your predictions? (Note 4.C)
Here, the micro-environment is sheltered: it is far from the lumen of the intestine, and net fluid flow is from the pancreatic cells towards the intestine.
Would you expect to see many mitotic figures? Would you predict any structural specialization to shield dividing cells from the luminal contents? (Note 4.D)

The bronchus

The epithelium that lines the major airways, the trachea and bronchi, is *pseudostratified* (Fig. 4.13). There are three major cell types present, two of which extend from basal lamina to lumen, the third being much smaller and resting on the basal lamina. One cell type we have already met, the mucus-producing goblet cell.

The environment faced by this epithelium is air, already warmed and moistened by its passage through the nose, pharynx and larynx. The air can contain dusts, pollen, bacteria, cigarette smoke and many other pollutants. As one might predict from the presence of so many goblet cells, the free surface of the epithelium is coated with a thin layer of mucus, which here has two functions. In heavy exercise in a dry environment, the air in the trachea and larger bronchi may not be quite saturated with water vapour. The mucus layer protects the cell surface from drying out by slowing down the loss of water to the incoming air. Next, the mucus is sticky, and airborne particles are collected on it like flies on flypaper, particularly at places where the airways branch. Obviously the trapped particles must be cleared somehow. This is achieved by propelling the mucus up towards the throat, where it is either swallowed or spat out. The lining of the major airways is thus coated with a layer of mucus that is carried slowly and steadily upwards, like a moving

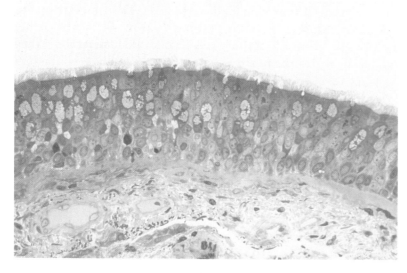

Fig. 4.13 Epithelium lining the human bronchus. (Courtesy, Dr J. Bertram.) (a) Light micrograph of plastic-embedded tissue (cf. Fig. 4.2). Toluidine blue. (× 385.)

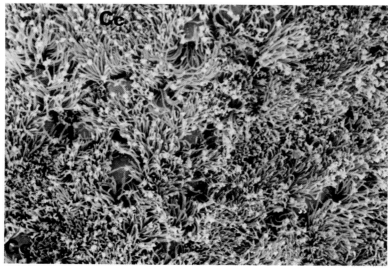

(b) Scanning EM of luminal surface, showing cilia and several goblet cells (× 2550.)

carpet, clearing the incoming air of particles. It seems that the goblet cells in the surface epithelium do not produce enough mucus to meet all conditions.

What other specialization of the epithelium might you predict to increase possible mucus production? (Note 4.E)

The mechanism for propelling this mucus layer is the beating of millions of cilia, and the cells bearing these cilia on their luminal surface make up nearly half the cells present (Figs 4.13b, 4.14). Seen alive under the microscope, this synchronized beating of cilia over a large sheet of epithelium is remarkable. Cilia are much larger than microvilli (Figs 4.13, 4.11), and have a characteristic structure with nine sets of paired microtubules surrounding two central single microtubules, all running parallel down the length of the cilium. Ciliary action is quite powerful, and the cilia are firmly attached by basal bodies to the feltwork of microfilaments at the cell apex, which are in turn associated with the intermediate junction and desmosomes.

What structure is likely to provide the means of communication between adjacent cells that is essential for

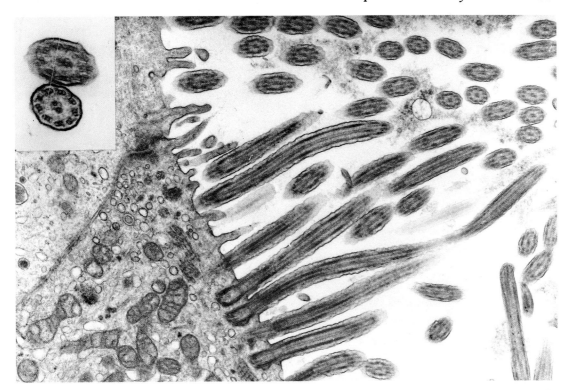

Fig. 4.14   Cilia in human bronchial epithelium. (×25 500.) Inset, the "9 + 2" arrangement of microtubules within the cilium. (× 62 000.) (Courtesy, Dr J.D. Henderson.)

the synchronized beating of cilia? (Note 4.F)
Ciliary action clearly requires energy. What cell component would you expect to find in high yield near the luminal end of ciliated cells? (Note 4.G)
The small rounded basal cells lie next to the basal lamina, well below the level of the tight junctions linking ciliated and goblet cells. They are thus largely protected from the external environment.
What would you predict to be their function? (Note 4.H)

The urinary bladder

Loss of water from the body is potentially lethal in animals like us that live in air. When waste materials have to be excreted, the solution in which they are carried, the urine, is concentrated to retain water. The epithelial layer facing the urine must therefore prevent two exchange processes: the diffusion of waste materials into the body from the urine, and the passage of water molecules out of the body into the urine by osmosis. Not surprisingly, it is a stratified epithelium: it is of the type called *transitional* (Fig. 4.15). It is a waterproof epithelium facing an aqueous environment. It is far from clear how this waterproofing is achieved. We do know that the outer membrane of the surface cells is thick and strangely rigid-looking, while other membranes of the same cells, and those of cells deeper in the epithelium, appear normal (Fig. 4.15b). This membrane appears to be synthesized in the surface cells and transported in flattened vesicles to be inserted into the outer cell membrane (Hicks *et al.*; 1974).

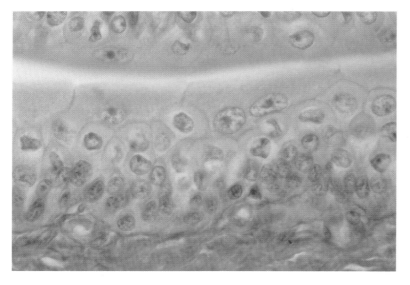

Fig. 4.15   Transitional epithelium lining the urinary bladder.
(a) Light micrograph. Carmine. (× 505.)

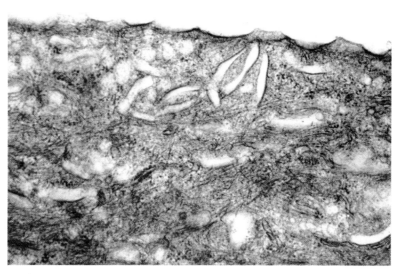

(b) TEM of the cell membrane facing the lumen. (× 30 000.) (Courtesy, Mr C. Yeo.)

In addition to survival in a fairly hostile environment, this epithelium must adapt to large volume changes within the bladder. This is done partly by folding of the epithelium in the empty bladder, and partly by changes in the cell attachments, which permit the cells to form many layers in the empty bladder while thinning out to 3–5 cells thick on stretching.

Would you expect cell replacement to be rapid? Where in this epithelium would you look for mitoses? (Note 4.1)

The oesophagus

This strong, muscular tube carries food and drink from the pharynx to the stomach. It is designed for rapid transport, and food undergoes no significant change during transit. In the resting state the oesophagus is closed, with its walls separated only by a thin film of fluid and mucus. At intervals, a wide range of objects comes past, some hard, some soft,

Fig. 4.16 Epithelium lining the oesophagus. H & E. (× 385.)

some liquid, hot tea, cold icecream, half an apple. The epithelium has evolved to withstand wear and tear rather than to secrete and absorb. Not surprisingly, it is *stratified.* (Fig. 4.16), and the surface cells are flattened.

Mitoses are fairly frequent, implying that surface cells are often rubbed off by passing food. Predictably, the dividing cells are segregated next to the basal lamina, in the most favourable micro-environment available. In every layer above this, the cells are firmly anchored to their neighbours by scores of desmosomes and the cytoplasm is full of bundles of microfilaments. If the cells shrink during histological processing, their cell membranes will draw apart, except at the desmosomes. This gives them a curious prickly appearance, as if they were connected to their neighbours by many spines of cytoplasm: hence the layer of cells between the basal layer and the superficial, flattened cells is called the *stratum spinosum.* As the cells age and get pushed slowly towards the surface, they die and flatten forming the squames that line the lumen.

This is not a waterproof epithelium, as loss of water to the lumen is of no consequence to the body. Mucus is present in the lumen, and helps to protect the epithelium against abrasion.

What structural modification might you predict that would allow the secretion of mucus on to a thick, stratified epithelium? (Note 4.J)

The skin

At first glance, the epidermis or epithelial layer of the skin (Fig. 4.17) closely resembles the oesophagus. It is a *stratified squamous epithelium*, with a basal layer concerned with cell division and a stratum spinosum above it. It is different only in being waterproofed. Water lost through the skin is lost completely from the body. The area of skin is so great that uncontrolled loss would restrict us to a very narrow range of environments.

Waterproofing is achieved in part by filling the flattened surface cells with a special protein, keratin. This is synthesized by a layer of cells between the spinosum and the flattened cells, the *stratum granulosum.* The granules are the first deposits of a precursor of keratin, keratohyalin, surrounded by free ribosomes. As keratohyalin builds up in the

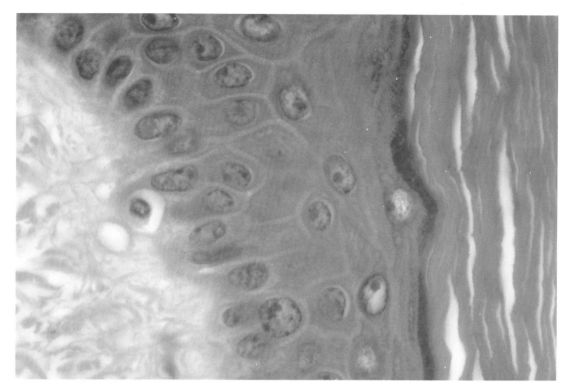

Fig. 4.17   The epidermis, with underlying connective tissue to the left. H & E. (× 1375.)

cytoplasm and becomes converted to keratin, the cells die and become homogeneous and translucent. This layer of dead, keratinized cells is remarkably important to our survival. Removal of the surface layers of skin, in burns or scalds involving a large area, for instance, may place the patient's life at risk through water loss and electrolyte imbalance.

Summary

You should now be able to recognize epithelium in a tissue section, and describe it using the standard classification. More important than that, you should have a framework by which to examine its structure and deduce something about its functions and the nature of the external environment it faces.

The layer of cells that marks us off from the outside world changes in structure and function from one place to another with surprising abruptness. Where a mucus gland enters the oesophagus, where the oesophagus enters the stomach, we see major changes occuring within one or two cells. Epithelia also change with time, responding to changes in the environment they face. Probably the best documented series of changes is that in bronchial epithelium in chronic exposure to cigarette smoke; it can change to a stratified squamous one. A breach in an epithelium is called an ulcer, a site of uncontrolled exchange between the tissues and the external environment, and a potential channel for invasion by micro-organisms.

Further reading

Moog, F. (1981). The lining of the small intestine. *Scient. Am.* **245:3**, 116–125. A concise, well-illustrated description of an absorptive epithelium.

Satir, P. (1974). How cilia move. *Scient. Am.* **231:4**, 44–52. The structure and beating of cilia analyzed.

Staehelin, L.A. and Hull, B.E. (1978). Junctions between living cells. *Scient. Am.* **238:5**, 140–152. A very clear review of junctions of all types.

References

Cereijido, M., Robbins, E.S., Dolan, W.J., Rotunno, C.A. and Sabatini, D.D. (1978). Polarized epithelial cells in culture and tight junctions. *J. Cell Biol.* **77**, 853–880.

Hicks, R.M., Ketterer, B. and Warren, R.C. (1974). The ultrastructure and chemistry of the luminal plasma membrane of the mammalian urinary bladder: a structure with low permeability to water and ions. *Phil. Trans. R. Soc. Lond. (Biol.)* **268**, 23–38.

Loewenstein, W.R. and Rose, B. (1978). Calcium in (junctional) intercellular communication and a thought on its behaviour in intracellular communication. *Ann. N.Y. Acad. Sci.* **307**, 285–307.

5

Connective tissue: the spaces in between

In the previous chapter, we have seen how the body is sheathed by a complete and continuous layer of cells, which controls the entry to and loss from the body of water, salts, food materials and waste products. This control at the frontiers of the body allows the creation of a constant and regulated environment inside, permitting the cells which live in the spaces between the epithelial surfaces the freedom to grow, differentiate and move about without uncontrolled fluctuations in the composition of the fluid that surrounds them. This freedom permits the development of many different types of cell, often highly specialized in appearance and function. Some collaborate to surround selected spaces within the body, creating volumes of water with different characteristics, while others accept the general environment of the body's internal spaces.

The immediate and obvious difference between an epithelium and the tissues beneath it is the size of the extracellular space (Fig. 4.2). In an epithelium, it is difficult to see any extracellular space: it comprises about 5% of the total volume of most epithelia. Beneath the epithelium, however, cells may appear widely separated from one another, and the extracellular space will often form 50–70% of the tissue. This space is not just fluid. It contains many fibres and other structures manufactured by cells.

Some of the cell types that occupy the body's interior are described later in separate chapters. The cells of the blood; the muscle cells that produce movement; the cells responsible for the skeletal elements, cartilage and bone, that help to translate the contraction of muscle into effective movement; the body's defences against invading micro-organisms and other foreign substances; the cells responsible for regulation and communication, the nerve cells and hormone producers: all these will receive separate attention in later chapters. But if we remove these highly specialized cells from the interior of the body, we are left with what is called connective tissue. It is almost impossible to describe adequately its distribution in the body. It surrounds all the specialized cell groups mentioned above, often extending between individual cells and compartments. It underlies every epithelium and, where the epithelium is thrown into complex folds and patterns, it extends between the folds (Fig. 4.12). It contributes to practically every tissue and organ in the body.

The basic components of connective tissue are the same throughout the body, though the relative amounts of these components, and hence the tissue's functional characteristics, may vary considerably from place to place. The three major elements making up connective tissue are the extracellular fluid, the cells that normally live in that fluid and the extracellular structures made by those cells.

The extracellular fluid

The cells that occupy the extracellular or tissue fluid take from it their nutrients and pass out into it their waste products. It is clear that the continued survival of the cells of connective tissue requires the renewal of the tissue fluid. This is achieved by exchange with the fluid component of blood. Without for the present describing the circulatory system, we need some picture of capillaries in order to understand the extracellular fluid.

Renewal of the extracellular fluid

Capillaries are narrow, cylindrical tubes, with internal diameters in the range of 5–8 μm. The wall of the tube consists of a single cell, surrounding the central lumen with a thin rim of cytoplasm. Numbers of such cells, end to end, make up the long, thin capillary. Outside them lies a basal lamina.
What is the composition of a basal lamina? What are its likely functions? (Note 5.A)
   Inside the capillary flows blood, which consists of a fluid component, the plasma, and suspended cells. The plasma contains many ions and other small molecules, and also larger molecules, the plasma proteins. The pressure within the capillary is higher at its arterial end than at its venous end, and this pressure difference drives the blood along the capillary. The capillary wall is effectively permeable to water, ions and small molecules, but retains the plasma proteins and cells. The tissue fluid is a clear, aqueous solution, similar in composition to the plasma but with a lower protein concentration. At the arterial end of the capillary, hydrostatic pressure is higher inside the capillary than in the surrounding fluid, and water is forced out through the capillary wall, taking dissolved salts and other small molecules with it. As the blood travels on down the capillary, two processes take place: the hydrostatic pressure drops steadily, and the proteins, which are unable to pass through the capillary wall, become increasingly concentrated, exerting an increasing osmotic effect. As a result, water passes from the tissue fluid back into the capillary at its venous end (Fig. 5.1).

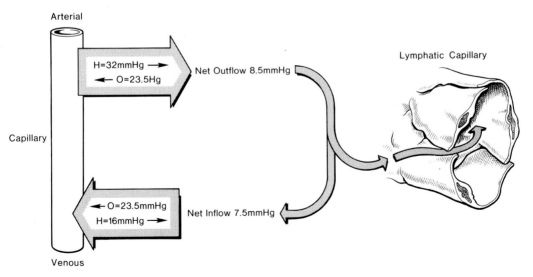

Fig. 5.1   The movement of fluid across the wall of a capillary. H = hydrostatic pressure: O = osmotic pressure.

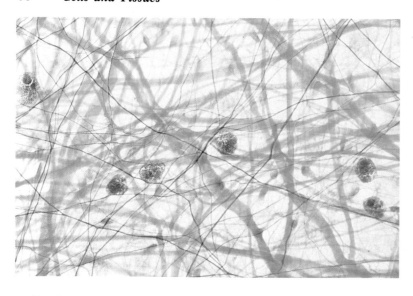

Fig. 5.2 A spread of connective tissue from the mesentary of the rat. Orcein and light green. (× 400.)

By this mechanism, the extracellular fluid of connective tissue is constantly renewed. Nutrients and dissolved oxygen diffuse into the tissue fluid and waste products, including carbon dioxide, are removed: large local variations in composition of the tissue fluid are prevented.

Removal of proteins from the extracellular fluid

Note the important part played by the plasma proteins in the renewal of the extracellular fluid: the return of fluid to the venous end of the capillary depends in part on the osmotic effect they exert. If proteins accumulate in the extracellular fluid, they cannot pass the capillary wall into the blood, and their osmotic effect opposes that of the plasma proteins, leading to retention of water in the extracellular spaces. Yet proteins appear in the extracellular fluid, since some plasma proteins escape from the capillaries and cells inevitably die, liberating debris, while the extracellular structures produced by them also break down.

The body has two mechanisms to prevent the gradual accumulation of proteins in the extracellular fluid — one cellular, the other an alternative route back into the blood stream. The cellular mechanism consists of the activity of macrophages, which take debris of all sorts into phagocytic vacuoles inside their cytoplasm. Digestion of debris and soluble proteins then takes place within the cell.

The second mechanism involves the provision of a second set of capillaries, of basically similar construction to those that contain blood, but of wider diameter: the terminal lymphatic capillaries are blind sacs lined with endothelial cells that overlap each other. Unlike the endothelial cells of blood capillaries, they are not joined to each other by tight junctions. The basal lamina is interrupted, and many endothelial cells seem to be anchored to neighbouring structures by fine fibres. The effect is to produce a valvular mechanism. When pressure in the surrounding fluid rises, pushing apart the cells and fibres of the extracellular space, the anchored endothelial cells are also pulled apart and fluid can freely enter the capillary lumen. When local movements compress surrounding tissues, the overlapping endothelial cells prevent leakage from the lumen into the extracellular spaces. Instead, the fluid is forced up into larger lymphatic capillaries. They in turn lead into still larger lymphatics, which have valves, so that movements of the body result in the slow flow

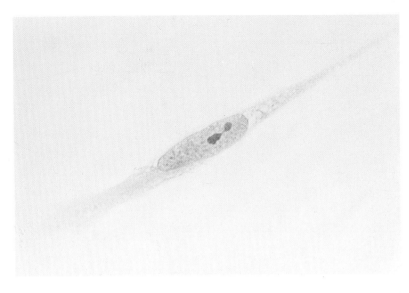

Fig. 5.3   A single fibroblast in tissue culture: before fixation it was moving towards the lower left corner of the picture. Iron haematoxylin. (× 1020.)

of tissue fluid up the lymphatics. This fluid passes through elaborate filters to be finally returned to the blood stream (p. 106).

The extracellular fluid, then, is constantly renewed, and its protein content is carefully controlled to a low level. This is the fluid environment for the cells of connective tissue. In many sites in the body, this fluid is also the source of oxygen and nutrients for nearby epithelial cells, and for the other, more specialized, cells such as muscle, which occupy the interior of the body.

The cells of connective tissue

Connective tissue can be examined, free from other cellular elements, in subcutaneous tissue spread in thin sheets, suitable for microscopy (Fig. 5.2). Such a preparation includes all the formed elements of connective tissue: the cells, the extracellular collagen and elastic fibres and the complex sugar-containing molecules, glycosaminoglycans (GAGs), which make up the structureless "ground substance". The components that stain readily are the cells and the extracellular collagen and elastic fibres; the fluid itself and the GAGs are not stained by the dye combinations in common use. In such a spread, the commonest cell type is the fibroblast, the cell responsible for the manufacture of the extracellular fibres and the GAGs. Next in order comes the macrophage, the remover of extracellular debris. Fat cells may be present in very large numbers in obese animals; their number varies with position in the body as well as with nutritional state. Finally, the mast cell occurs sporadically. In addition, other cell types may be found, having entered the connective tissues from the blood stream, particularly if there is some minor infection or tissue damage. Since these cells are present in far higher concentration in the blood, they will be considered in detail in Chapters 7, 8 and 15.

Fibroblasts

These cells synthesize most of the collagen, elastin and GAGs in the extracellular space. Fibroblasts are capable of movement, but spend most of the time relatively stationary. They are typically elongated cells (Fig. 5.3), making contact with neighbouring fibroblasts

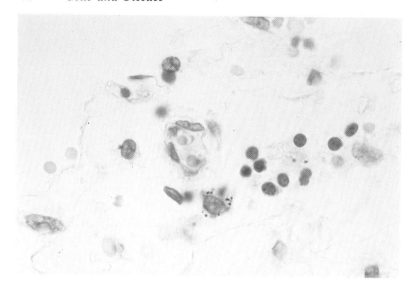

Fig. 5.4   Cells in loose connective tissue from human lung. One cell just below centre has black granules of phagocytosed carbon: it is a macrophage. H & E. (×1020.)

by irregular processes of cytoplasm. Most of their surface membrane, however, is bathed by extracellular fluid. The nucleus is usually rather oval, and may be cigar-shaped. Fibroblasts vary considerably in their degree of activity. Some, surrounded by dense strands of collagen in places where the extracellular components only need replacing very slowly, are almost inactive. Others, in the edges of healing wounds, for instance, are synthesizing new extracellular materials very rapidly. Fibroblasts grow readily in tissue culture, and many observations have been made on them in this artificial environment.

Collagen is a protein, and its synthesis by fibroblasts has been much studied. It is secreted, as fast as it is synthesized, in the form of procollagen, with a molecular weight of about 112 000, which contains small amounts of carbohydrate. At or near the outer surface of the cell membrane, this molecule is converted to the alpha chains of collagen, which cross-link outside the cell to produce collagen itself. Thus, though the initial synthesis of collagen is intracellular, the final assembly of the molecule takes place in the extracellular fluid.

Given these facts about collagen synthesis, what would you expect an actively synthetic fibroblast to look like by transmission electron microscopy? (Note 5.B)

The orientation of collagen fibres is crucial to their function in connective tissue, and it is still much debated how the orientation of the fibroblast affects the final direction in which newly synthesized collagen runs. Some claim that the long axis of the cell determines the direction of new collagen; others, that collagen is oriented in response to factors in the extracellular environment, and that the collagen may impose a particular orientation on the cell. Certainly, fibroblasts will orient themselves precisely in the absence of collagen. Fibroblasts grown in a drop of culture medium on a slide will be randomly arranged, but if the drop is spread in one direction by means of two fine needles, producing an elongated drop instead of a hemispherical one, the fibroblasts will, within 24 hours, align themselves to the long axis of the fluid.

The elastic fibres of connective tissue have two components, a central core of the protein, elastin, and microfibrils surrounding it which are also protein, but different in composition (Fig. 5.9b). Like collagen, elastin is secreted by fibroblasts as a precursor, protoelastin, and it is in the extracellular space that this is assembled into elastin, surrounded by

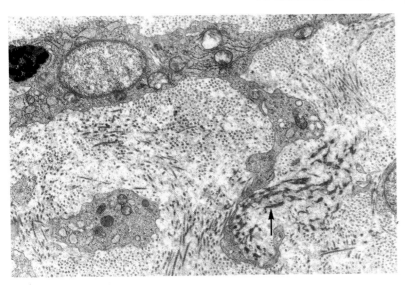

Fig. 5.5 TEM of macrophage in the uterine tissues of an ovariectomized rat. Note the irregular, deeply stained collagen about to be phagocytosed (arrow). (×10 000.) (Courtesy, Miss C. Lunam.)

microfibrils, to make elastic fibres.

The third major product synthesized by fibroblasts is ground substance, jelly-like material consisting of glycosaminoglycans (GAGs). GAGs are polysaccharides that contain aminosugars.

Which organelles are likely to be concerned in the synthesis of GAGs? (Note 5.C)

Fibroblasts can be found in nearly every histological section. At the light microscope level, there is little to help one identify them with certainty, apart from their association with collagen fibres and their elongated, often cigar-shaped nuclei (Fig. 10.1).

Macrophages

The macrophage is the cell responsible for removing proteins and other debris from the extracellular fluid, and digesting them. In some tissues, it can become a fixed and stationary cell, but in general it is actively moving. In cine films or videotapes of connective tissue, macrophages almost seem to be boiling. Their surfaces are always moving, with a series of cytoplasmic processes forming and sinking away again, while the whole cell moves about continually. The general shape of the cell is more rounded than that of the fibroblast, but fixation and sectioning may produce almost any outline from such a variable cell.

In spreads of connective tissue (Fig. 5.2), it is not easy to be sure which nuclei belong to macrophages. In general, their nuclei are rounder and slightly smaller. In sections, the very inhomogeneous cytoplasm of the macrophage may allow one to recognize it: the presence of phagocytosed material in varying stages of digestion is responsible for this appearance. *What appearance would you predict for a macrophage when viewed with the transmission electron microscope? (Note 5.D)*

The simplest way to identify macrophages with certainty is to place some easily recognizable foreign substance, such as particles of carbon, in the extracellular fluid: those cells that accumulate the particles are, by definition, macrophages (Fig. 5.4). But macrophages do not only digest debris. If tissues or organs need remodelling, extracellular materials such as collagen can be removed by macrophages. (Fig. 5.5). How they "recognize" which collagen fibres to digest is a mystery.

(a)   (b)

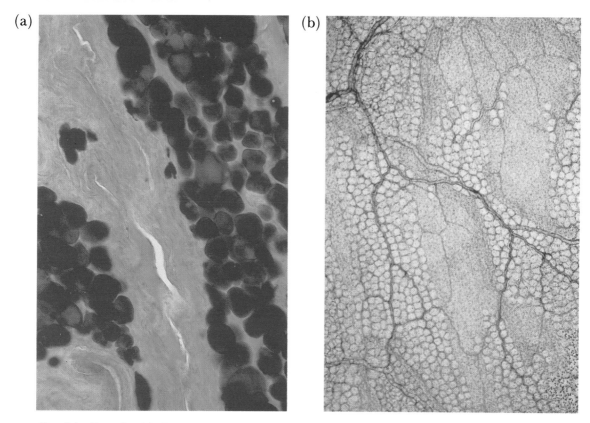

Fig. 5.6   Fat cells. (a) Frozen section with lipid preserved. Oil Red O. (× 127.) (b) Spread from mesentery, with lipids removed by solvents used in processing tissue. Silver impregnation. (× 40.)

The macrophages of connective tissue can travel. They appear in many tissues, and often have different names. In blood, they are called monocytes; in the liver, they are Kupffer cells; in lymph nodes and spleen, they are reticulo-endothelial cells. In every site, they are capable of dividing, and also of synthesizing new lysosomes, so that they provide a flexible and efficient system for removing unwanted materials, and particularly proteins, from the extracellular fluid.

Fat cells

Many cells in the body store lipids as a source of energy for future use: in these cells, the fat is usually in the form of many small droplets in the cytoplasm (Fig. 3.15). In some sites in the body, however, cells in connective tissue become specialized for fat storage, and in this case the droplets coalesce to a single, large droplet, which often becomes very big in cellular terms, achieving a diameter of over 100 μm and reducing the cytoplasm to a small rim around the droplet, in which the nucleus lies, a thin, flattened crescent.
What will such a cell look like with the light microscope and standard methods of specimen preparation? (Note 5.E)
With the transmission electron microscope, the cytoplasm contains the usual organelles — mitochondria, RER, Golgi apparatus, and a few free ribosomes. There may also be small lipid droplets which have not yet coalesced with the main drop.
Would you expect these droplets to be separated from the cytoplasm by membrane? (Note 5.F).

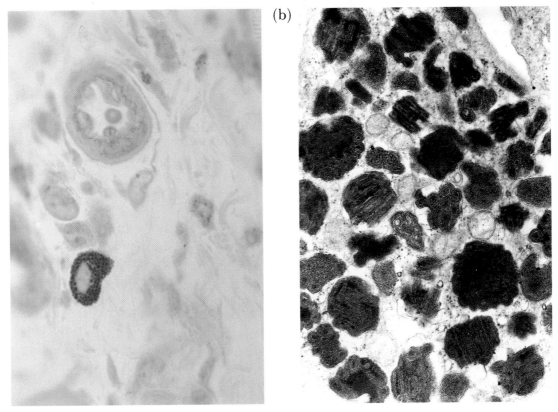

Fig. 5.7 Mast cells. (a) Cell filled with granules, staining metachromatically. Toluidine blue. (× 1015.) (b) TEM of cytoplasm of human mast cell. (× 25 000.) (Courtesy, Dr M. Haynes.)

Fat cells are more active than they look, with lipid continually being mobilized and replaced: this activity presumably reflects the alternating phases of feeding and fasting through the day and night.

Fat is liquid at body temperature, and collections of fat cells are capable of taking up any shape within wide limits, under pressure from surrounding tissues. We shall see in Chapter 10 that fat pads commonly occur near and even in joints, as one of the specializations of connective tissue to accommodate movements. Fat is also rather poorly vascularized by comparison with connective tissue in general. A thick layer of fat cells offers some small degree of insulation against heat transfer, and we shall see such a layer in the skin (Chapter 6).

Students often have difficulty initially in distinguishing fat cells in standard histological sections from capillaries and small lymphatic vessels. Fat cells usually occur in clusters in connective tissue, and typically have a hexagonal outline, while capillaries do not occur in similar clusters, have a more circular profile, and a lumen smaller than the fat droplet, with a nucleus that is less flattened and compressed (Fig. 5.6). Lymphatic capillaries have a larger diameter in general than blood capillaries, and may have more than one nucleus per profile (Fig. 8.1).

Mast cells

Scattered through the connective tissues of the body are rounded cells with a round, central nucleus and a cytoplasm filled with granules; with the light microscope, these granules may mask the nucleus completely. Some dyes show metachromasia in association with these granules (Figs 5.2, 5.7a).

What does metachromasia mean? (Note 5.G)

These are mast cells. It may be difficult in standard sections to preserve them adequately, but, if they are prevented from rupturing and spilling out their granules during histological processing, there is no mistaking them for any other cell type; the granules themselves appear to be soluble in many aqueous fixatives, which is a further reason why this very common cell is so seldom seen in class sections.

Seen with the TEM, the mast cell is the ultimate in intermittent secretion. The granules, which are surrounded by membrane, appear to fill the cytoplasm, and it is difficult to find the synthetic apparatus which produced them. RER, Golgi apparatus and mitochondria do exist, on careful inspection (Fig. 5.7b). The granules contain heparin, a potent anti-coagulant, which is a GAG. Histamine is also present, a substance that, when released, causes other mast cells to secrete their granules, and has powerful effects on blood-vessels locally, increasing the permeability of capillaries and small venules and increasing blood flow. Mast cells secrete as one component of the defence mechanisms of the body against injury and infection; we shall study them in more detail in Chapter 15.

For the present, it is enough to know that mast cells synthesize and store signalling substances for immediate release in case of tissue damage. Their curious staining reactions are due to the high content of sulphated GAGs in their granules.

These, then, are the four types of cell that regularly inhabit the connective tissues of the body — fibroblast, macrophage, fat cell and mast cell.

Extracellular materials of connective tissue

All connective tissue contains three classes of material in the extracellular space — collagen fibres, elastic fibres and GAGs. Within each class there are significantly different subclasses: the relative amounts of each and their arrangement differ from one position to another in the body. In some areas, collagen fibres are very plentiful, almost to the exclusion of extracellular fluid and the other classes of material: such areas are referred to as dense connective tissue. In other places in the body, extracellular fluid and GAGs predominate, with few bundles of collagen or elastic fibres. Histologists have attempted to set up elaborate classifications of connective tissue based on these variations, but it is important to realize that the groups they have named are part of a continuous spectrum, not clearly defined entities.

The important thing is to be able to recognize the three classes of extracellular material in connective tissue, and to understand the function of each. It should then be possible to look at sections of any connective tissue and to work out its characteristics.

Collagen fibres

Collagen fibres are clearly seen in Fig. 5.2, as thick, wavy bundles passing through the extracellular fluid in many directions. At lower magnification in Fig. 2.7, collagen in varying densities can be seen throughout the connective tissues. These fibres have great tensile strength: in other words, they resist stretching forces, though, like a rope, they can

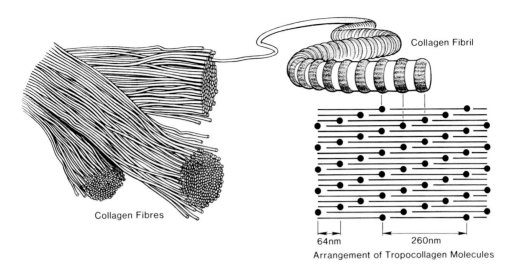

Fig. 5.8   *The structure of collagen fibres.*

be coiled or crumpled. Collagen fibres are not only very widely distributed in our bodies, they occur throughout most of the animal kingdom, wherever tensile strength is needed.

What characteristics are necessary in a collection of protein molecules to produce a tough, unstretchable fibre? Many aminoacids are quite long chains of atoms, linking to form polypeptides which can assume a number of different structural patterns. Collagen is designed for resistance to elongation at the aminoacid level. The two principal aminoacids are hydroxylysine, a very short molecule, and hydroxyproline, a five-membered ring with four carbon atoms and one of nitrogen, which together form polypeptides with little or no possibility of rearrangement to permit elongation.

The aminoacids are assembled into a polypeptide chain, four of which link together to form a coiled helix with a molecular weight of about 110 000, called an α-chain. Three such helices then combine to form a single molecule of tropocollagen, with a molecular weight of about 336 000 consisting of some 3000 aminoacids. Tropocollagen molecules are thin structures 260 nm in length and about 1·5 nm wide.

How large would such molecules be at the magnification of Fig. 5.9b? (Note 5.H)

We are still a long way from the wavy bundles of fibres of Fig. 5.2. The molecules of tropocollagen polymerize, head to tail, and the resulting threads cross-link with other threads in a characteristic pattern, so that the heads of the molecules in one thread are displaced by one quarter of the length of the molecule, relative to the neighbouring thread. Single collagen fibrils contain variable numbers of such threads, but, because of the characteristic linking between threads, all have transverse bands clearly demonstrable by electron microscopy, which occur every 64 nm — one quarter of the length of the tropocollagen molecule (Fig. 5.8). The thick bundles of Fig. 5.2 are groups of fibrils running parallel, to form collagen fibres.

Synthesis takes place within the fibroblast up to the formation of α-chains: these are secreted in modified form by the cell as procollagen, and are converted to α-chains by an enzyme at the cell surface. The formation of tropocollagen from these chains and the polymerization of tropocollagen into collagen fibres occur outside the cell.

(a)

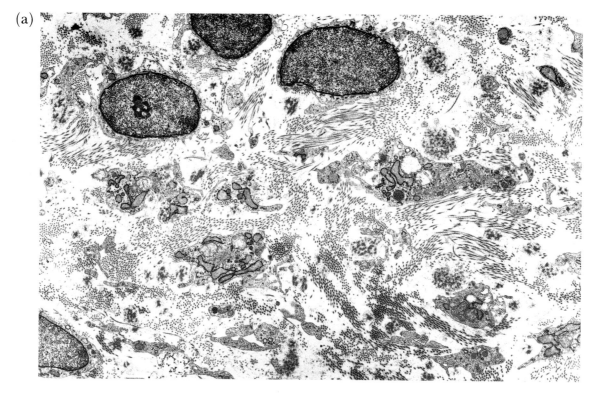

(b)

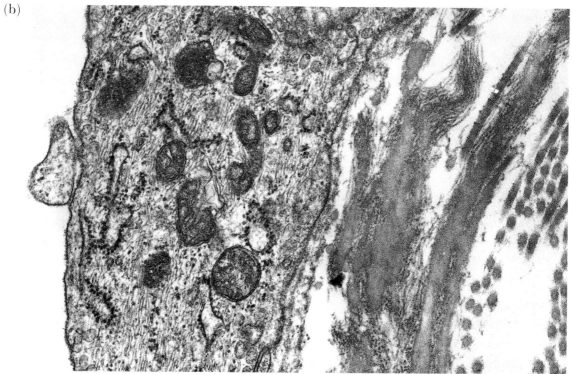

Fig. 5.9   TEM of collagen and elastic fibres. (a) Survey view of connective tissue with fibroblasts and fibres. (× 3500.) (b) Fibroblast cytoplasm on left; elastic fibres, centre; collagen, right. (× 95 000.) (Courtesy, Dr J. Fanning.)

Collagen has many uses in the body. Fine fibres, composed of relatively few fibrils but still with the characteristic cross-striations, occur in basal laminae as a delicate meshwork. These adsorb silver salts in certain conditions, unlike thick fibres of collagen, and were called reticular fibres, in the belief that they were somehow different; the name is still used for very fine collagen. In loose connective tissue, collagen fibres run in every direction, like a feltwork. Such tissue can be deformed in any direction until sufficient bundles of collagen become aligned to resist further movement. In sheets of interwoven fibres, collagen forms the tough, white fascia which surrounds muscles. With the fibres all aligned and parallel, collagen makes up ligaments and tendons which have a quite remarkable ability to resist stretching. The eyeball consists of many layers of collagen fibres, arranged in sheets at right angles to each other, and the fluid inside is maintained at higher pressure than the surrounding tissues, like a blown-up football: in this way the distance of retina from cornea is kept constant, yet the eye is tough and absorbs the energy of a direct blow like a kicked football, returning immediately to its original shape. We shall see in Chapter 12 that peripheral nerves are very fine cytoplasmic processes of nerve cells, sometimes as much as a metre long. Such a process offers almost no resistance to tugging forces. Each process has its own sheath of collagen, however, which helps to prevent its easy rupture. Scar tissue is also collagen.

Remember, collagen resists stretching, but can easily be crumpled. If resistance to deforming forces is needed, collagen fibres have to be embedded in some matrix which prevents them from being crumpled. In Chapter 10 we shall see that collagen is the basis of cartilage and bone, which each have different mechanisms for preventing crumpling.

Many cell types synthesize collagen, though in relatively small amounts: the epithelial cells that make basal laminae are an example. Whereas collagen may be 1–3% of the total protein synthesized in such cells, it forms over 50% of the total in fibroblasts.

Elastic fibres

Returning again to Fig. 5.2, the collagen forms thick, wavy bundles which do not branch. Elastic fibres, by contrast, are thin, single and branching. These fibres can be stretched, each fibre extending by up to 50% of its original length. When the stretching force is removed, these fibres spring back to their former length. In loose connective tissue, elastic fibres permit some deformation of the tissue and return it to its original position afterwards. Unlike collagen, they do not have a high tensile strength and are easily ruptured by excessive force. In connective tissue, the collagen and elastic fibres are finely balanced so that the collagen resists forces that would be strong enough to rupture the elastic fibres.

Seen in the TEM, elastic fibres are surprisingly disorganized in appearance (Fig. 5.9). A central, shapeless mass is surrounded by peripheral microtubules arranged in bundles and running longitudinally down the fibre. The shapeless material is the protein, elastin, which is synthesized in fibroblasts and secreted as a precursor, tropoelastin. In the extracellular space, these molecules polymerize to form elastin. The microtubules are composed of a different protein, and seem to function by providing orientation to the otherwise shapeless elastin.

Elastic fibres occur quite widely in the body, providing the springiness of the walls of arteries and the recoil of the lungs which follows the muscular effort of breathing in. One of the changes that occur in the body in old age is a loss of elastic fibres and GAGs, with a whole series of consequences to the appearance and function of the skin, the lungs, the arteries and so on. Like collagen, elastic fibres may be incorporated into cartilage; they have no place in the more rigid tissue, bone.

Elastic fibres, then, occur throughout loose connective tissue, and in much higher concentration in a few other sites. They are not specifically stained with haematoxylin and eosin (H & E).

What staining methods have you seen which demonstrate them at the light microscope level? (Note 5.I)
With the light microscope, elastic fibres appear smaller and finer than collagen; with the TEM, however, elastic fibres are clearly wider in diameter.
Can you explain this? (Note 5.J)

Glycosaminoglycans

The microscopists of 100 years ago became convinced that the extracellular spaces were not just clear fluid with collagen and elastic fibres in it, but contained some material without structure which provided a jelly-like consistency to the fluid. This they called ground substance. Since it had no demonstrable organization, the word amorphous was often tacked on to its name.

We now know that amorphous ground substance is made up of a number of related species of molecules called glycosaminoglycans (GAGs): these are complex polysaccharides that contain aminosugars. There are many different GAGs, and their relative amounts vary from place to place in the body: the two most widely distributed are hyaluronic acid and chondroitin sulphate, each a name for a group of related molecules.

Hyaluronic acid has a molecular weight between 200 000 and 500 000, and forms a coiled, approximately spherical mass which occupies a surprisingly large volume. It is negatively charged at positions all down the length of the molecule, and binds cations, particularly sodium. In consequence, such a molecule holds, effectively bound to it, a large amount of water. This diffuse, hydrated molecule imparts great viscosity to the extracellular fluid, preventing the free flow of water, and the spread of large molecules by convection and diffusion. Invading micro-organisms meet the resistance to spread provided by this component of the ground substance, and some synthesize and release hyaluronidase to assist their penetration through tissue.

Chondroitin sulphate is also a large, coiled molecule, but this time highly sulphated. Like hyaluronic acid, its repeating, negatively charged groups enable it to bind sodium and other cations in large amounts, and thus to retain water.

So, when we stand up, all our extracellular fluid does not flow down into our feet. The collagen and elastic fibres do not just run through a watery fluid, but are surrounded and entangled with these large, hydrated GAGs, which help to retain them in position and give some restraint to their crumpling. We have to modify our ideas about the renewal of extracellular fluid: Fig. 5.1 is clearly a gross over-simplification, since much of the extracellular water is not free to flow, but is held, as if it were frozen, by the GAGs. It seems that there are well-defined channels through the extracellular space, between the GAGs and around the fibres, along which exchange of extracellular fluid occurs. The passage of molecules in solution down these channels is not free and easy, since these channels may be narrow, and are lined by highly charged GAGs to which dissolved materials may become adsorbed, even if they do not form ionic bonds with them.

We shall see in Chapters 9 and 10 how the resistance to compression imparted by GAGs can be used in the formation of cartilage, and how the viscosity due to GAGs contributes to the functioning of joints.

Summary

The spaces in between the major cell groups of the body are quite complex and highly structured, providing varying degrees of physical support and accommodating changes in size and shape. Most important of all, throughout the body these spaces carry the capillaries, and deliver to surrounding cell groups the nutrients and oxygen derived from the blood and dissolved in the extracellular fluid.

Further reading

Casley-Smith, J.R. (1980). The fine structure and functioning of tissue channels and lymphatics. *Lymphology* **12**, 177–183. A brief summary in a symposium on lymphatics.

Guyton, A.G. (1981). "Textbook of Medical Physiology", 6th Ed. W.B. Saunders, Philadelphia. Chapters 30 and 31 cover well the formation and exchange of extracellular fluid and the lymphatics.

Kleinman, H.K., Klebe, R.J. and Martin, C.R. (1981). Role of collagenous matrices in the adhesion and growth of cells. *J. Cell Biol.* **88**, 473–485. A review of the types of collagen and their synthesis, leading on to the adhesion of cells to collagen and its effects on cell behaviour.

Ross, R. (1973). The elastic fibre. A review. *J. Histochem. Cytochem.* **21**, 199–208. A full account of structure, well illustrated.

Sandberg, L.B., Soskel, N.T. and Leslie, J.G. (1981). Elastin structure, biosynthesis, and relation to disease states. *New Engl. J. Med.* **304**, 566–579. A shorter review with few electron micrographs, but summarizing the contributions made to pathological conditions by elastin failure.

Wasserman, S.I. (1979). The mast cell and the inflammatory response. *In* "The Mast Cell: Its Role in Health and Disease", Pepys, J. and Edwards, A.M. (Eds). Pitman Medical, London. A short introduction to the mast cells' role in inflammation.

White, A., Handler, P., Smith, E.L., Hill, R.L. and Lehman, I.R. (1978). "Principles of Biochemistry", 6th Ed. McGraw-Hill, New York. Read pp. 1146–1158 for an account of GAGs.

6
How to look at a section

We are quite familiar with those things we can see and touch. The further we go from this level to smaller and smaller structures, the greater the imaginative leap we need to make to "see" objects in their surroundings. Always start from naked-eye observations, even when looking at a section on a slide; progress through increasing levels of magnification, and use the highest powers of the light microscope last in your sequence of looking, only if they are really necessary.

The sequence of observations

It is surprising how much information can be obtained from naked-eye viewing of a microscope slide. Is the specimen a section of solid tissue, or uniformly spread across the slide, like a smear of blood cells? What shape is the section? Is it small or large, uniform or obviously varying from one part to another? If it is a section of solid tissue, are the edges suspiciously straight and at right angles to each other, suggesting a block cut out of a larger piece of tissue? Is it a hollow tube? Are there several separate pieces of tissue?

If you begin by examining a small bunch of cells, chosen more or less at random, using the × 40 objective, you make it unnecessarily difficult for yourself to work out what they are like and what they might be doing, since you have no bridges with structures with which you are familiar, and have no idea of the organization of tissues around the small area you are looking at.

Figure 6.1 is a naked-eye view of a section. It is oblong, with one long side apparently covered by a wavy, densely stained layer, and the other three sides rather straight, as if they might have been cut out with a scalpel.

A quick look with a hand-lens often fills out the information gained naked-eye. Not many of us carry such lenses, Sherlock Holmes style, but the eyepiece of the microscope is a good alternative. Remove one eyepiece, turn it round so that the section lies where your eye would normally be, and place your eye near the end that sat in the tube of the microscope. With a bit of adjustment you will see the section suitably magnified. Check your preliminary observations. Are those likely to be cut edges? What does the layer on the fourth side of the section look like? Is the rest of the tissue homogeneous, or are there big differences in cellularity and staining from one place to another?

Now, and only now, are you ready to use the lowest power of your microscope.

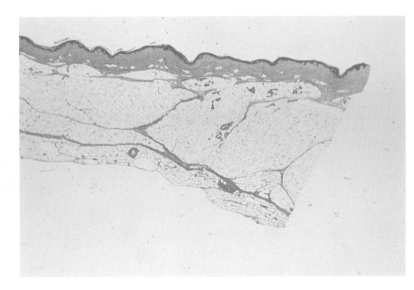

Fig. 6.1 A naked-eye view of a section. H & E. (× 10.)

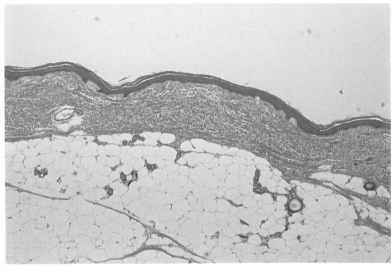

Fig. 6.2 One part of the section seen in Fig. 6.1. H & E. (× 62.)

Remember, any increase in magnification inevitably means that you are looking at a smaller area of section. Figure 6.2 shows some details of the same section, seen with an objective × 4 and an eyepiece × 10. Work around the edges first.

What conclusions can you come to about the layer covering the section? What sort of tissue is this? (Note 6.A)

Look next at the other three surfaces.

Do they appear to be natural surfaces, or cuts through tissue? (Note 6.B)

Now examine the whole section quickly.

Is it homogeneous? Is there any identifiable epithelium visible, apart from that at the surface of the section? (Note 6.C)

There will almost always be loose connective tissue present.

Are there any special features of the connective tissue, or any special cell groups within it that do not appear to be epithelial? (Note 6.D)

Now that you have some idea of the distribution of different cell groups across the slide,

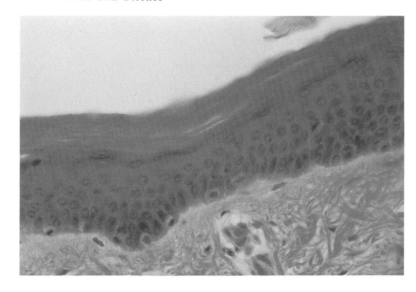

Fig. 6.3   The epithelial surface of the section seen in Figs 6.1 and 6.2. H & E. (× 400.)

take each in turn, look at it with the × 10 objective, checking particular points with the × 40 objective but returning at once to the × 10 lens. Never search large areas of slide with the × 40 or × 100 objectives: use these lenses only for answering specific questions about cells or structures you have already seen at lower magnification.

It may be satisfying to identify cells synthesizing proteins for intermittent secretion from their characteristic structural features, or cells that are synthesizing steroid hormones, and so on, at a final magnification of × 1000; but this does not tell you if you are looking at a small group of cells in the wall of a tube, or the major cell type in a solid organ.

Examining a section

Through even the early stages of examining a section with a hand lens, try to decide whether or not any of the surfaces are epithelial. The cellularity and polarization of epithelia usually give sufficient clues. It is often more difficult to spot glandular epithelium, which does not line an obvious surface. It helps to be familiar with the patterns in which glandular tissue is organized (Fig. 4.5).

We have already seen that capillaries are carried to within a very short distance of epithelial cells in the loose connective tissue beneath the basal lamina. It is reasonably easy to visualize the change needed in this arrangement of connective tissue to bring a blood supply to a simple, tubular gland. With increasing complexity of organization of the gland, the need to surround each epithelial element with a sheath of connective tissue carrying blood-vessels remains just as important, and produces a three-dimensional pattern which is much more difficult to picture. Remember that a section may cut through a gland at any angle, and will only show you a thin slice of tissue, in which epithelial cells, connective tissue and blood-vessels may be arranged in a bewildering sequence. Remember also that such a section will not always cut through the lumen of a gland or its duct.

Returning to our section at higher magnification, the epithelial surface is shown in Fig. 6.3.

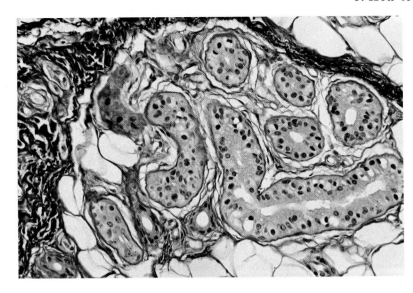

Fig. 6.4 Section through a gland, surrounded by connective tissue. H & E. (× 230.)

What type of epithelium is this? What deductions can you make about the environment at the surface of this epithelial layer? (Note 6.E)
Looking below this epithelial sheet, other epithelial elements, probably glands, can be seen forming two distinct types of structure, and possibly three. Figure 6.4 shows one of these.
How would you classify this type of gland? (Note 6.F)
The ducts of these glands can occasionally be seen, winding their way to the surface, though it is almost impossible to follow one for any distance in a single section.

These glands have an obvious lumen, but the second class of epithelial structures is at first sight more puzzling (Fig. 6.5). It appears to be a cylinder of cells with a solid, rod-like centre in which individual cells cannot be distinguished. These structures communicate with the surface. They may be cut through at any angle, but they never appear coiled. Attached to the side of this structure is sometimes seen another epithelial area (Fig. 6.6), the cells of which seem large and rather moth-eaten, as if some constituent had been dissolved out in histological processing.
What might this component be? (Note 6.G)
There is no clear lumen, but the cells lying centrally near the communication of this gland with the main epithelial downgrowth from the surface look large and irregular, and the nuclei are not well defined. These cells are dying and breaking down. This is a sebaceous gland, opening into a hair follicle.
What is the term applied to the mechanism of secretion of a sebaceous gland? (Note 6.H)
This is a section of skin, with a stratified, squamous epithelium at the surface, showing keratinization, and sweat glands and hair follicles.

Structure and function in skin

We will now look at the histology of skin in more detail, trying to mobilize the knowledge already available to us in order to link the histological structure of skin to its functions.

(a) 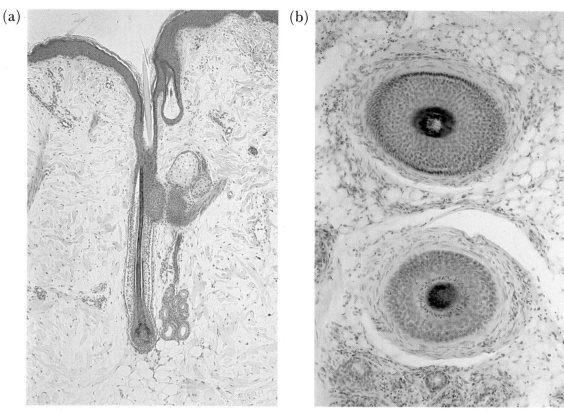 (b)

Fig. 6.5 Sections through an epithelial structure below the epidermis. (a) Longitudinal view, showing communication with surface H & E. (× 60.) (b) Transverse sections. Iron haematoxylin. (× 100.)

The epidermis

The surface epithelium of skin is the epidermis; the underlying connective tissue is the dermis. Look now at the epidermis on the back of your hand. It is dry and smooth, in spite of the extracellular fluid just beneath it: in other words, it is waterproof. If you place a drop of water on the back of your hand, it remains there, without sinking in. If you rub the skin gently, very fine scales come off the surface, though you will need good sight and lighting to see them. If you now look at the palm of your hand, the epidermis feels thicker, and may be in thickened callouses if you have been doing manual work. One other difference will be obvious if you have been in the sun lately — the back of your hand will be brown with sunburn, but the palm will not.

The structural basis for these observations can be seen in Figs 6.3 and 6.7. The epithelium in Fig. 6.3. is stratified, but relatively thin; that in Fig. 6.7 is also stratified, but very much thicker. The former is from the forearm, the latter from the sole. In a stratified epithelium, only the cells of the basal layer lie on the basal lamina. Here, they have the best access to nutrients and are in the most protected position, being furthest from the varying outside environment.

What is the most likely function of these cells in the stratum basale? (Note 6.I)

The cells lying superficial to the stratum basale appear to be linked to their neighbours at

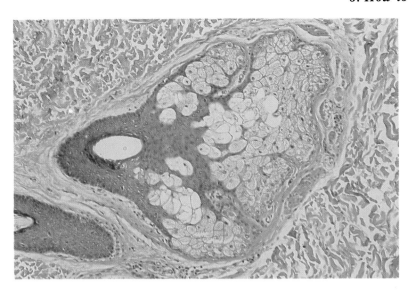

Fig. 6.6   Gland attached to side of structure illustrated in Fig. 6.5. In Fig. 6.5a, this gland is halfway down the structure, on the right. H & E. (× 100.)

many points on their surfaces, having an appearance which gives the layer its name, the stratum spinosum.

What would you expect to see at these points of contact? What intracellular component might you expect to find converging on these structures? (Note 6.J)

Superficial to the stratum spinosum, a thin layer of cells contains very obvious, deeply stained granules (Fig. 4.17). The exact nature of these granules is not well understood, but the material is called keratohyalin, and their appearance signals the conversion of the living cells of the stratum spinosum to the dead, flattened plates of keratin seen in the more superficial layers. The layer superficial to the stratum granulosum is clear and translucent, and no nuclei are visible: it is as if the cells had become converted into a homogeneous, non-living material. With the TEM, these cells have none of the normal intracellular organelles, though desmosomes are still clearly visible. The microfilaments have been embedded in a matrix of keratin, and no longer stand out as they do in watery cytoplasm. This layer is the stratum lucidum.

Finally, above the stratum lucidum lies the stratum corneum, composed entirely of flattened, keratin-filled ghosts of cells. This is the layer responsible for waterproofing the skin, and it is this layer that provides the dust-like plaques of keratin that are rubbed off by abrasion.

Epidermal thickness. Comparing Fig. 6.3 with Fig. 6.7, the greater thickness of the stratum corneum on the sole is obvious, but note that the stratum spinosum is much thicker also. It is common knowledge that repeated abrasion results in a thicker epidermis — "the horny-handed sons of toil", and so on. Is this change in the environment the explanation for the differences between the epidermis of forearm and palm? Curiously enough, it is not the full story. Figure 6.8 shows comparable micrographs of the skin of the sole of the foot and the leg from a foetus at the end of pregnancy. Although skin in both sites is cushioned in the same fluid environment before birth, the epidermis of the sole is clearly thicker. The factors in the epidermis that determine its thickness, either before birth or in response to use, are quite unknown. The rate of cell division in the stratum basale

*Fig. 6.7   Thick epidermis from the sole of the foot.
H & E. (× 40.)*

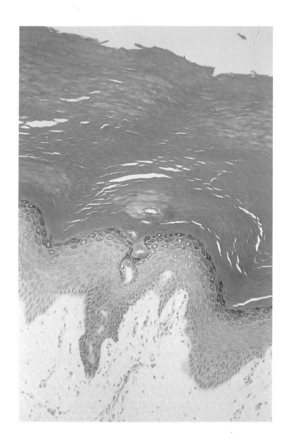

increases with more rapid removal of surface layers, but that alone need not necessarily result in increased thickness: it might, for instance, produce a rapid rate of passage of cells across an epidermis which is the normal thickness. To increase thickness, the time spent by each cell in the epidermis as well as the rate of cell division must be precisely regulated.

Pigmentation. We have looked at the structural basis for most of our observations about living epidermis, with the exception of its characteristic, brown colour. This is due to granules of the black pigment, melanin, which are normally present in many cells of the stratum spinosum and more superficial layers (Fig. 6.3). Curiously enough, this pigment is not synthesized in these cells but in a separate cell type, the melanocyte. These have their embryological origin in the neural crest and migrate into the skin in the foetus. They are rather flattened cells lying on the basal lamina amongst the cells of the stratum basale, and, seen from the skin surface, they have many long processes which spread out, branching and climbing between the basal and spinous cells. Melanocytes synthesize melanin, but contain very little of it. They appear to secrete the pigment granules from their processes, and the surrounding cells take them in. The number of melanocytes is fairly constant in human skin, whether from a person with white or with heavily pigmented skin: the difference in skin colour seems to be a function of the activity of melanocytes rather than their number. On exposure to sunlight, increased synthetic activity produces more melanin granules. Since these are rapidly transferred to cells of the spinous layer, they are ultimately lost when these cells are shed from the surface of the skin. To remain brown, a fair-skinned person must receive repeated exposure to sunlight.

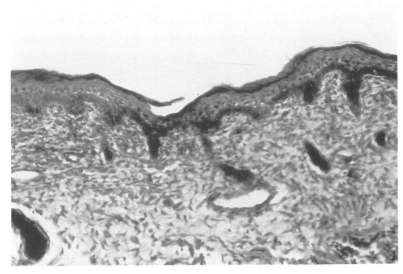

Fig. 6.8 Skin from a foetus in the 6th month of intrauterine life. Azocarmine. (× 200.)

(a) Sole of foot.

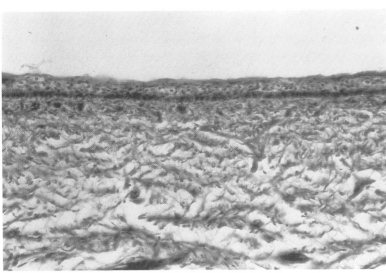

(b) Side of lower leg.

Dividing cells are very sensitive to ultraviolet radiation, which is present in sunlight, and the superficial layers of the epidermis are relatively thin and translucent. The presence of melanin in the epidermis helps to screen the basal cells from UV radiation, an overdose of which can cause sunburn, and, in the long term, skin cancers.

The dermis

Beneath the epidermis, the transition to dermis is very obvious, with wide spaces between cells (Fig. 6.2). What can we discover about the dermis by examining our own skin, and what predictions can we make about its histological structure?

Dermal connective tissue. First, the dermis is a definable layer distinguishable from the connective tissue deeper in the body. If you gently pinch up the skin of your forearm, it is clear that the fold is thicker than epidermis alone, and the tissue in the fold moves over the

(a) (b)

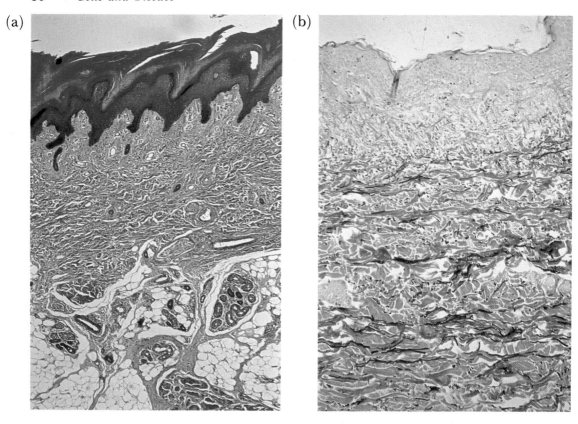

Fig. 6.9   Connective tissue of the dermis. (a) Mallory's trichrome. (× 40.) (b) Verhoeff's & VG. (× 40.)

underlying muscles of your arm. So, between the dermis and the deeper tissues lies a plane which permits a certain amount of movement.

Looking at this movement more closely, if you place a fingertip on the skin of the back of your hand, you can move finger and skin from side to side, but only within certain limits. Beyond this point, movement gets increasingly difficult, and ultimately painful. The same is true of movement up and down the long axis of your hand.

What tissue component resists stretching? How can this component be arranged to permit initial movement, but to limit excessive movement? (Note 6.K)

Now pinch up a fold of skin again, and let it go. It springs back to the normal, flat position.

What tissue component is likely to be responsible for this springiness of the dermis? (Note 6.L)

The organization of the dermis (Fig. 6.9) must provide a structural basis for these observations. Note how dense the collagen fibres are in the dermis, and how sparse they are in the layer immediately deep to it, permitting the movement of skin on deeper structures which we have already noted. In some parts of the body, the dermis is firmly attached to deeper structures: look at the creases on the palms of your hands, for instance.

How might this tethering of the dermis be achieved? (Note 6.M)

Remember that fat is fluid at body temperature, and that it is packaged in many separate fat cells (Fig. 5.6). A layer of fat, such as that seen in the dermis, can adapt readily to changes in shape produced by pressure or movement. This layer has one other possible function. Fat cells have a relatively low metabolic rate, and thus a sparse blood supply, and a layer of fat cells may act as an insulating layer against heat transfer by conduction, due to its limited blood flow.

Apart from these mechanical properties of the dermis, we are all familiar with the changes in skin colour produced by blushing and by turning pale. These happen so rapidly and reversibly that they must be due to changes in the amount of blood flowing through the dermis, rather than in the density of blood-vessels.

What is blood flow through the dermis like in someone very hot? What if you are very cold? (Note 6.N)
Clearly, the dermis must contain sufficient blood-vessels to provide the epidermis also with oxygen and nutrients. Its network of vessels is in addition extensive enough to produce considerable heat loss when the vessels are open and blood is flowing through them fast, and sufficiently well-regulated to close down to minimal levels of blood flow if the body requires to conserve heat.

You may wonder how the very thick epidermis of palms and soles obtains sufficient oxygen and nutrients. Thin epidermis has a fairly level, flat junction with dermis, but, in thicker epidermis, this epidermal–dermal interface becomes very irregular. Each extension of dermis up into the epidermis contains blood-vessels, and this is partly a device to bring capillaries nearer to the more superficial layers of living cells in a very thick epithelium (Fig. 6.7).

In summary, then, the dense feltwork of collagen fibres in the dermis, with its additional elastic fibres, provides structural strength to the skin, while the layer of fat and the loose connective tissue below the dermis allow the skin to move over deeper structures. This plane of loose tissue allows us to skin an animal, while the dense layer of dermal collagen, when suitably treated, becomes leather.

There remain the specialized epithelial elements situated in the dermis, the sweat glands and the hair follicles.

Sweat glands. We are all familiar with sweating, the production of a salty, watery fluid on the surface of the skin, which cools the body by evaporation. We also tend to sweat when nervous, particular on the palms of the hands and soles of the feet. Remembering that the epidermis is a waterproof layer, it is plain that the production of sweat must be a function of specialized glands, opening on to the epithelial surface at numerous sites. These pores are clearly visible on the skin, if you look carefully at it with a hand lens.

The secretory cells of the sweat glands (Fig. 6.4) are of two types. The clear cells are small, and may not extend to the main lumen of the tubule. They do face on to small extensions of this lumen which lie between the cells, however, and this surface of the cell has an extraordinarily irregular shape, with many folds of cell membrane. A number of large mitochondria lie just below this membrane. Apart from this specialization, these cells have very few organelles, with remarkably little endoplasmic reticulum and only a small, insignificant Golgi apparatus. The cytoplasm contains large amounts of glycogen.

What function would you expect such a cell to have? (Note 6.O)
The second cell type is the dark cell. This is a larger cell, and its apical membranes form the margins of the tubule lumen. It has a lot of RER, a prominent Golgi apparatus, and many membrane-bound vesicles clustered near the cell apex, containing material that gives histochemical staining reactions for mucoproteins and mucopolysaccharides.

What might be the function of this cell? (Note 6.P)

The duct of the sweat gland winds up towards the epidermis, losing its specialized epithelial lining as it enters the epidermis: here, it is simply a sinuous channel through the epithelial cells and the stratum corneum.

One other cell type may be seen in sweat glands, the myo-epithelial cell. In a number of epithelia, including that of sweat glands, specialized, spider-shaped cells lie between epithelium and basal lamina, with their processes round the epithelial tube. In section, it is difficult to get a good impression of their size and shape. They contain contractile proteins (see Chapter 9), and are thought to assist in emptying the gland by contracting and squeezing the lumen.

Sweat glands produce anti-bacterial substances and a secretion with a characteristic odour. It is likely that the dark cells contribute these materials to the watery sweat produced by the clear cells. Certainly the sweat glands in places like the axilla possess a higher proportion of dark cells.

Hair follicles. Let us assemble what we know about hair. It varies very much in length and thickness from the scalp to the forearm, and is missing altogether from the palms and soles (Fig. 6.7). It is continually growing, and hairs often come right out. Especially on the scalp, hair is somewhat greasy. In "goose-pimples", the hairs stand upright, when you are cold or frightened.

The hair follicle (Fig. 6.5) is a straight, tubular gland, with a mass of connective tissue carrying capillaries projecting up into an indentation at its base. Epithelial cells around this dermal papilla divide rapidly and continuously, and produce the keratin-containing cells of the very centre of the growing hair. As this medullary rod grows upwards towards the surface of the skin, epithelial cells around the circumference of the follicle divide and mature, contributing keratin-containing cells to the cortex of the hair. The epithelial cells of the region of follicle nearest the surface clearly provide the outermost layer of the emerging hair. The result is a cylindrical rod of keratinized cells, growing out from the opening of the hair follicle. This gland produces and "secretes" keratinized cells.

In just the same way as in surface epidermis, melanocytes synthesize the pigment, melanin, and pass it to developing cells in the equivalent of the stratum spinosum, for inclusion in the hair itself. The greasy secretion on the hair is provided by the sebaceous glands (Fig. 6.6), one of which opens into each hair follicle just below the surface of the skin. The cells of sebaceous glands synthesize the oily secretion, which is anti-bacterial and assists in waterproofing the hair.

What appearance would you predict for such a cell, seen by TEM? (Note 6.Q)

This is an example of holocrine secretion.

What does this mean? (Note 6.R)

How do hairs "stand on end" when you are frightened or cold? A small band of muscle, the arrector pili, is attached at one end to the side of the hair follicle, at or below its midpoint, and at the other to the connective tissue just below the epidermis. When this muscle shortens, the hair follicle, which normally lies at an angle to the surface, is brought more upright: at the epidermis end, the muscle produces a tiny pit or dimple when it contracts, giving the characteristic appearance of "goose-pimples". In relatively hairless mammals such as ourselves, this is not very effective, but it has two functions. The first is to trap a thick layer of stationary air next to the skin, which acts as an insulator against heat loss.

The second is to make the animal appear larger, and hence more frightening to an opponent.

Summary of the functions of skin

Let us briefly review the functions we have noted for the skin, revising in each case the histological structures responsible for that function.

Protection

(i)   Against mechanical abrasion. The stratified epithelium loses cells at the surface, replacing them continually from below.
(ii)   Against water loss. The stratum corneum is waterproof.
(iii) Against invading micro-organisms. The dark cells of the sweat glands and the sebaceous cells produce anti-bacterial secretions, protecting the glands of the skin, while the structure of epidermis itself prevents bacterial penetration.
(iv)   Against UV light. The pigment melanin acts as an absorbent screen.
(v)   Against stretching and deforming forces. The tough, collagenous dermis supports the epidermis, and moves on the underlying tissues, absorbing energy as it deforms. The elastic fibres assist in this, and return the skin to its original position when the deforming force ceases.

Temperature regulation

(i)   Variations in blood flow through the dermis increase or reduce heat loss.
(ii)   The poorly vascularized fatty layer in the dermis reduces heat conduction from the body to the surface of the skin.
(iii) The hairs, standing erect on contraction of the arrector pili muscle of each follicle, trap a thicker layer of stationary air, reducing heat loss.
(iv)   Evaporation of sweat causes cooling.

One final function of skin must be mentioned, though it will not be discussed at length here. As the interface between us and the outer environment, the skin provides us with a great deal of information, about where it is being touched, about pressure, heat and cold, and pain. The skin has an extensive sensory nerve supply.

Review of progress

Always try to link the structures you see down the microscope with what you know already of the anatomy and physiology of the organ concerned. Start the examination of every slide by looking at it naked-eye, and build a low-magnification picture of the tissue before looking at selected individual cells at higher power. Identify epithelia first, then examine the distribution and organization of connective tissue. If other, specialized cell groups occur, deal with them next.

Whether you are trying to identify an unknown tissue or studying a known one, use your intelligence while you observe. The structures you see are organized to carry out particular tasks efficiently, and the links between structure and function are often relatively easy to spot.

Further reading

Bloom, W. and Fawcett, D.W. (1975). "A Textbook of Histology", 10th Ed. W.E. Saunders, Philadelphia. Try the chapter on skin, pp. 563–597.

Elias, P.M. (1981). Epidermal lipids, membranes, and keratinization. *Int. J. Dermatol.* **20**, 1–19. A review of recent work on the waterproofing of the epidermis.

7
The blood

It is clear that, in a multicellular organism above a certain size, the cells deep inside the body do not have direct access to external oxygen or nutrients, and cannot readily get rid of their carbon dioxide and waste products. The best they can do is to draw what they need from and excrete their waste into the extracellular fluid. The main function of the circulating fluid in the blood-vessels is the renewal of the extracellular fluid in the body, and thus the maintenance of the stability of the internal environment. In its circulation, the blood acts as the agent of the extracellular fluid in communicating with the external environment, to carry out the exchange of oxygen and carbon dioxide, the collection of nutrients and the excretion of waste materials.

But there is a limit to the solubility in water of many of the materials required by the extracellular fluid. A wide spectrum of carrier proteins has evolved, which permits the blood to transport greater amounts than are directly soluble. This pattern applies to many substances including aminoacids, hormones, and in particular oxygen. Now proteins exert a given osmotic pressure in solution, as well as influencing other factors such as viscosity, which are crucial to the flow of plasma down blood-vessels, and there is a limit to the concentration of carrier proteins compatible with functional requirements. So we find that haemoglobin, the carrier protein for oxygen, is not in solution in the plasma, but is packaged within the cell membranes of red blood cells, where it can have little effect on plasma osmolarity and viscosity, while remaining in dynamic equilibrium with the free oxygen in solution in the plasma.

Apart from the carriage of materials to and from the extracellular spaces, the blood carries proteins and cells which together form a mobile reserve against invasion of the body by foreign cells, in particular micro-organisms. Such invasions may occur anywhere on the body's surfaces, and micro-organisms may be distributed by lymphatic or blood flow from their original beach-head to any site or tissue in the body. The circulating proteins with a defensive role include antibodies against foreign substances which have already been encountered by the body. The cells include a large, short-lived population, the polymorphonuclear leucocytes, supplied with ready-made primary lysosomes and capable of immediate attack on invading micro-organisms. In addition, a smaller group of monocytes circulates: these are also phagocytic cells, but differ from the polymorphs in their ability to divide and to synthesize new lysosomal materials. These monocytes provide a second, more flexible reserve of defensive cells, but cannot give the rapid response of the polymorphs to bacterial invasion. The third group of circulating cells is the various types of lymphocyte, the cells responsible for the immune response of the body to foreign substances and cells. These defensive proteins and cells normally remain within the circulating blood. They can leave it for the extracellular spaces in great amounts when tissue damage triggers the inflammatory response (Chapter 15).

Finally, the blood stream must flow without hindrance through normal, uninjured vessels, but its flow must be stopped where injury to the vessel wall, and in particular a breach in the wall, occurs. The blood contains an elaborate mechanism for producing an insoluble clot to prevent continuing flow in damaged vessels.

So blood is a tissue with varied functions and considerable complexity. While classified as one of the connective tissues of the body, it contains neither collagen nor elastic fibres, and its cells remain separate and circulating. It is an essential component of connective tissues everywhere, exchanging with the extracellular fluid.
How is this renewal of extracellular fluid brought about? (Note 7.A)
Unlike every other tissue, blood consists of cells and proteins that are synthesized elsewhere in the body, and are removed from the blood prior to their destruction.

Blood plasma

This straw-coloured, aqueous solution forms only about 55% of the blood by volume, the cells comprising the other 45%. Its composition is closely controlled, so that its osmolarity, its pH and the levels of many of its constituents remain within strict limits. The plasma proteins, as we have seen, play an important part in the exchange of fluid with the extracellular spaces. Of these, the albumins are the ones in highest concentration, at 35–50g/l, followed by the various globulins and then fibrinogen. These proteins are the only elements of plasma which are recognizable histologically: they can be precipitated by histological fixatives to produce in the blood-vessels an uniform, jelly-like material, stained faintly pink with eosin.

Carrier proteins are found amongst both the albumins and the globulins. Circulating antibodies are globulins. The third major group of plasma proteins is involved in the clotting of the blood when the endothelial lining of blood-vessels is damaged, and in restraining the clotting mechanism in the absence of such damage. In clotting, soluble fibrinogen is converted into insoluble fibrin, which forms a feltwork of fibres in which blood cells become enmeshed.

The albumins and fibrinogen are synthesized in the liver, and released directly into the blood stream. The antibody globulins are the products of scattered plasma cells in many tissues, often finding their ways through lymphatic vessels into the blood.

The cells of the blood

There are four groups of cells that normally circulate in the blood, and, in addition, fragments of cytoplasm called platelets. The vast majority of the cells are erythrocytes, or red blood cells, of which there are about 5×10^6 in every cubic millimetre of blood. The three types of white blood cells are the polymorphonuclear leucocytes (which can be further classified on the basis of their cytoplasmic granules), the lymphocytes (all of which look alike, but which include cells with at least two different life histories), and the

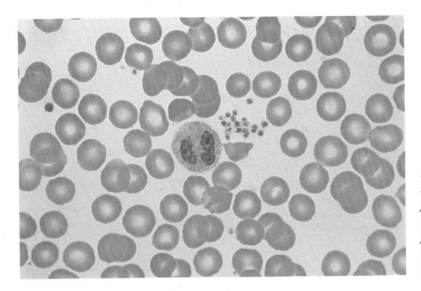

Fig. 7.1 Human blood smear, showing one neutrophil polymorphonuclear leucocyte, with a cluster of platelets to its right, surrounded by red blood cells. Leishman's stain. (× 1015.)

monocytes, which are the circulating form of the macrophage of connective tissues. The white cell count in an adult human is 5000–10000 cells per cubic millimetre of blood.

Erythrocytes

An erythrocyte (Fig. 7.1) is little more than a packet containing the protein, haemoglobin, surrounded by cell membrane. It has no nucleus, having lost it in the course of differentiation, no ability to synthesize protein or other macromolecules, and few recognizable organelles in its cytoplasm. Yet its size and shape are remarkably constant. It is a disc, thicker at the rim than in the middle, and its diameter is about 7·2 μm in dried films, about 8·5 μm in suspension. It appears to contain microfilaments, arranged to maintain its biconcave shape, but these are difficult to see in its cytoplasm, which is uniformly dense in transmission electron micrographs, due to the iron-containing haemoglobin. Although erythrocytes have the same, constant shape in blood samples, they can be deformed as they pass down capillaries, which are usually smaller in diameter than the red cell: on emerging into a venule, the erythrocytes revert to their normal shape.

Erythrocytes are very frequently seen in sections, providing a useful indication of the sizes of surrounding structures. They are the one cell type that does not normally pass out through the walls of capillaries, but remains in the blood stream. If they occur free in the tissues, this means that one or more blood-vessels have ruptured, either during life or, quite commonly, during the collection of the tissue for histological processing. So if erythrocytes are seen in a tube, that is a blood-vessel. The converse is not necessarily true: small capillaries may not contain an erythrocyte in the short length within a tissue section, while some material is fixed for histology by washing out the blood and perfusing fixative down the blood-vessels.

The haemoglobin of erythrocytes is like the carrier proteins of plasma in that it allows blood to carry much more oxygen in bound form than could ever go into solution directly. It would be impossible to maintain the osmolarity of plasma within reasonable limits if the haemoglobin were to circulate as a plasma protein: packaging it inside a cell membrane effectively removes it from the class of osmotically active substances in the blood. The biconcave cell shape is highly efficient for exchange with free oxygen in solution: the surface area is large, at about 140 μm^2, and no point within the cytoplasm is far from the surface membrane.

Given the lack of synthetic activity in the erythrocyte, one might expect it to be a short-lived cell. In fact it has a surprisingly long survival time in the circulation, of about 120 days.

Erythrocytes are produced in the bone marrow in post-natal life; in the foetus, the liver and spleen are additional sources. Old erythrocytes are removed from the circulation largely in the spleen, although macrophages in the liver and bone marrow may also take part in this process.

Polymorphonuclear leucocytes

These are quite small cells, little larger than erythrocytes (diam. 10–12 μm). They get their name from the very variable shape of the nucleus, which is often a series of small lobes linked by thin arms (Fig. 7.1). Such a nucleus contains mainly clumped chromatin; there is almost no disperse chromatin and no nucleolus.
What does this observation tell you about the activity of the cell? (Note 7.B)
This nucleus develops from a more typical spherical nucleus, with a much higher percentage of disperse chromatin. It follows that a cell with a bilobed nucleus is younger than one with 4 or 5 lobes.

These small, round cells have a cytoplasm filled with granules: in fact, they are classified on the basis of the staining reactions of these granules.

The neutrophil. By far the commonest polymorph is the neutrophil (Fig. 7.1). In man, this has very small granules, which appear cytochemically to be of two types. The first, forming the majority, is called a specific granule, and is secretory. It contains basic proteins called phagocytins, which are anti-bacterial. The second type of granule appears to be a large, primary lysosome. These cells are actively phagocytic and mobile. Curiously, for a cell with so many secretory granules and primary lysosomes, the cell has very little RER and only a small, centrally placed Golgi apparatus.
What do these observations suggest about the functions of the cell? (Note 7.C)

The eosinophil. The next group of polymorphs is the eosinophils (Fig. 7.2). These have only one class of granule, and this is bigger than those of the neutrophil. With the TEM (Fig. 7.3), these granules contain crystalline inclusions, while cytochemically they appear to be large, primary lysosomes.

Whereas the neutrophil is actively phagocytic, particularly liking bacteria, the eosinophil, lacking the secretory granules of anti-bacterial proteins, appears specialized for the removal of antigen–antibody complexes. Both cell types have very short lives in the blood, of about 5 days. Essentially, they are a constant, ready line of defence against invading micro-organisms. Incapable of either cell division or protein synthesis, they are fully equipped to immobilize, ingest and destroy bacteria (neutrophils) or ingest and destroy complexes of antibodies with antigens (eosinophils). They represent an expendable

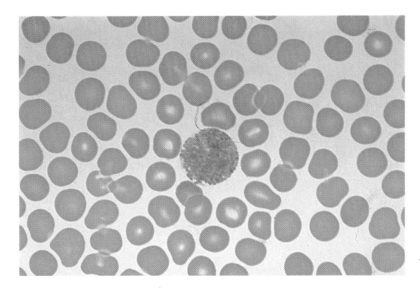

Fig. 7.2 Human blood smear with one eosinophil polymorphonuclear leucocyte. Leishman's stain. (× 1015.)

population, circulating and ready for immediate response. Though they are usually confined to the blood, they can migrate through the walls of small blood-vessels to reach sites of infection or damage. Pus consists of millions of dead and dying polymorphs.

Both neutrophils and eosinophils are produced in the bone marrow in very large numbers, and enormous reserves of nearly mature cells are available for release into the blood if needed. It has been calculated that there are 200 eosinophils in the marrow for every one circulating. Normally, there are about 5000 neutrophils per cubic millimetre of blood, but less than 500 eosinophils.

The presence of many neutrophils outside blood vessels strongly suggests infection with micro-organisms; the presence of many eosinophils in the tissues suggests allergy or hypersensitivity.

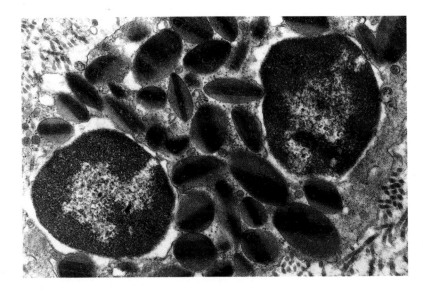

Fig. 7.3 TEM of eosinophil polymorphonuclear leucocyte showing crystalline inclusions in granules. (× 23 000.)

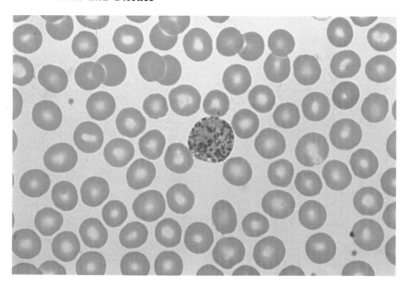

Fig. 7.4 Human blood smear, with basophil polymorpho-nuclear leucocyte. Leishman's stain. (× 1015.)

The basophil. The third class of polymorph is considerably rarer than the other two. The basophil has a cytoplasm full of large, basophilic granules (Fig. 7.4), which appear to be rather fragile, rupturing easily in common fixatives. These cells have many features in common with the mast cell of connective tissues. The granules contain heparin and histamine, are fragile, and produce metachromasia with certain dyes. Although it has been hotly debated in the past, it now seems that the basophil has no direct relationship to the mast cell: the basophil is produced in the bone marrow from precursor cells similar to those of neutrophils and eosinophils, while the mast cell arises in connective tissue. Both release granules in the early stages of inflammation, with similar effects (Chapter 15). The basophils appear to act as a mobile reserve for the mast cells, which tend to be relatively stationary.

Lymphocytes

These are also small, spherical cells, with a diameter about the same as that of erythrocytes.
What is the diameter of an erythrocyte? (Note 7.D)
Lymphocytes are common cells, making up about a quarter of all the white blood cells. Unlike the polymorphonuclear leucocytes the lymphocyte is nearly all nucleus, with only a thin rim of cytoplasm around it (Fig. 7.5). The nucleus is spherical or slightly indented on one side, and contains much condensed chromatin, while the cytoplasm is quite remarkable for its paucity of organelles. The lymphocyte does not look like an active cell, lacking the organelles associated with the synthesis of materials for export, and those to do with phagocytosis and lysosome formation.

Lymphocytes were mysterious cells up to about 20 years ago, with many fierce debates about their functions. They do not perform any very obvious activity, yet they are long-lived, and can be followed in the tissues by observation through transparent windows inserted surgically for periods of many days without changing or apparently doing anything. In fact they are the cells responsible for immune reactions against foreign proteins; they will be discussed in detail in the following chapter.

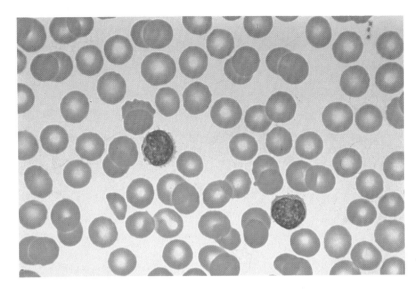

Fig. 7.5 Human blood smear with two small lymphocytes. Leishman's stain. (× 1015.)

As we shall see in the next chapter, the one term, lymphocyte, includes at least two cell types with different functions and histories. As their appearance suggests, they are inactive cells, each waiting for the appropriate stimulus — the arrival of one particular antigen in the tissue — before responding in its own specific fashion.

Monocytes

These are the largest of the cells in blood smears, with a diameter of 15–17 μm, but in suspension they are spherical and smaller, with diameters of 10–12 μm: it seems that they flatten out during the preparation of smears (Fig. 7.6). Monocytes can be confused with the largest lymphocytes: both have a spherical or indented nucleus, and a cytoplasm free from large, obviously stained granules. But, on closer examination, there are significant differences. The monocyte has a larger, more palely staining nucleus, with one or two

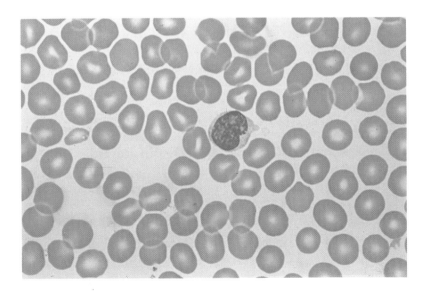

Fig. 7.6 Human blood smear, with one monocyte. Leishman's stain. (× 1015.)

nucleoli. The cytoplasm has small lengths of RER and many free ribosomes, and there are a number of membrane-bound granules about 400 nm in diameter which give positive cytochemical reactions for acid phosphatase.

How do you interpret these observations? (Note 7.E)

These cells are produced in the bone marrow, and form only about 5% of the circulating white cells: they leave the blood within a day or two to enter the tissues.

What are these cells called when they arrive in the tissues? (Note 7.F)

Unlike polymorphonuclear leucocytes, monocytes retain the ability to divide and to synthesize new lysosomes.

Platelets

These are visible on blood smears as tiny, eosin-stained specks with central, basophilic granules (Fig. 7.1). They are often disc-shaped, 2–3 μm in diameter, lacking a nucleus. By transmission electron microscopy (Fig. 7.8), they have a peripheral ring of microtubules and a central volume of cytoplasm containing glycogen and an occasional mitochondrion; in addition, they contain membrane-bound granules. Platelets are produced in the bone marrow from very large cells, the megakaryocytes, from which small fragments of cytoplasm break off into separate existence as platelets. They are concerned with the blocking of small defects in the walls of blood-vessels, and the initiation of clot formation. We shall examine them in action later in this chapter.

The blood summarized

To review briefly the composition of the blood, it is an aqueous solution in which many molecules are carried for distribution to the tissues, or for excretion, principally by the kidneys. It contains a spectrum of proteins, the main function of which is to bind active molecules, increasing the carrying capacity of the blood: one of these, haemoglobin, is present in such massive concentration that it is segregated in erythrocytes, but the rest are in solution. Apart from its functions in preserving homeostasis in the tissues, the blood carries a number of proteins and cell types concerned in protecting the tissues against invading micro-organisms and foreign proteins: the polymorphs and monocytes are actively phagocytic, the lymphocytes are involved in the immune response, as are the various immunoglobulins. These cells and proteins are capable of leaving the blood to enter the tissues at sites of invasion or damage. The third class of constituents of the blood is those involved in clotting when the walls of a blood-vessel are damaged, and those which normally prevent clotting in the absence of damage. Platelets, fibrinogen and the many other factors which control clotting are in this group.

At the capillaries, water, dissolved salts and other small molecules leave the blood to pass into the extracellular fluid, but the plasma proteins and cells do not. The plasma proteins exert sufficient osmotic pressure to assist in the return of water and solutes from the extracellular fluid at the venous end of the capillary.

One further point has been made about the cells and proteins of the blood. All are produced elsewhere. The polymorphs, monocytes, platelets, erythrocytes and many of the lymphocytes come from the bone marrow; other lymphoid organs produce the remaining lymphocytes. Most of the plasma proteins are synthesized in the liver, apart from some

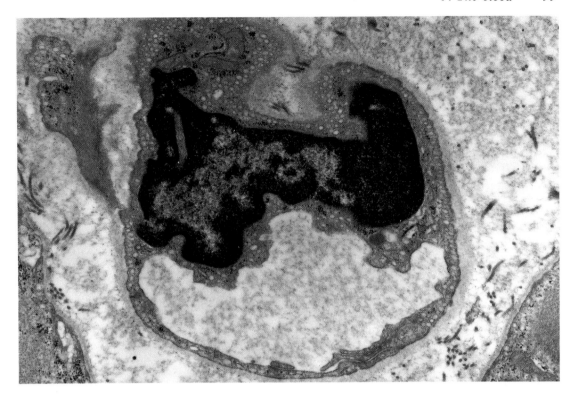

Fig. 7.7   TEM of capillary in human skeletal muscle. (× 13 500.) (Courtesy, Dr M. Haynes.)

immunoglobulins, which are produced in lymphoid tissues throughout the body. The blood is a specialized compartment in which its cells live for only part of their life span. Remember that for each type of blood cell, the circulating cells are only a fraction, often a tiny fraction, of the total population of that type of cell, and that one cannot tell by looking at them down the microscope how long they spend in the blood stream. The total number of polymorphs in the blood is in consequence a very crude index of the numbers entering and leaving at any given time.

Exchange between blood and tissues

The blood-vessels and the heart form a system of channels, closed off from the surrounding tissues by a continuous layer of cells, the endothelium, through which exchange takes place between the blood and the extracellular fluid. The capillary wall consists of the cytoplasm of endothelial cells surrounded by a basal lamina (Fig. 7.7). In the heart and the arteries and veins, the endothelium is surrounded by variable numbers of layers of cells, collagen and elastic fibres. In section, endothelial cells are thin and flattened, with flattened, crescentic nuclei: many of the capillaries of connective tissue appear on section to be a central space surrounded by the cytoplasm of a single endothelial cell, which is rolled like cigarette paper around a cigarette. Where the processes of cytoplasm meet, they are joined by tight junctions.

What does a tight junction look like by freeze fracture? What is believed to be its function? (Note 7.G)

Capillaries

These vary in diameter but are often about 5–8 μm, permitting the passage of erythrocytes and white cells as well as plasma. In sections, the nucleus of an endothelial cell is crescentic, with much condensed chromatin. The cytoplasm is a very thin layer around the lumen, with few mitochondria or lengths of RER, and frequently large numbers of pinocytotic vesicles.
What are these? How are they formed? (note 7.H).

Outside the capillary endothelial cells lies a basal lamina, similar to that underlying epithelia, and, outside that again, the extracellular spaces of connective tissue, with collagen and elastic fibres and glycosaminoglycans. There has been much debate about the transfer of water, electrocytes and other molecules across the capillary wall, some workers maintaining that transport in pinocytotic vesicles plays an important part, others arguing that minute clefts between the cytoplasm of adjacent endothelial cells provides a more likely pathway.

In some tissues, capillaries have distinct specializations of the endothelial cells, apparently to increase their permeability. These are called fenestrations (Figs 7.10, 3.9). At these sites the two cell membranes on luminal and outer surfaces of the endothelial cytoplasm unite to form a circular pore, when viewed by freeze fracture. Such pores occur in clusters on the capillary wall. Fenestrated capillaries are found in endocrine glands, in the absorbtive regions of the intestine and in the kidney glomerulus, all sites where increased permeability would be an advantage.

Pericytes. Capillaries are so delicate and thin that they can have almost no ability to contract: regulation of the flow of blood through a capillary bed is almost entirely by constriction or relaxation of the smallest arteries feeding into it. One possible mechanism for altering the diameter of capillaries locally does exist, however: it is the pericyte. This is an octopus-shaped cell, which can sometimes be seen closely attached to the outside of a capillary, with its processes wrapped around it. With the TEM, pericytes can be seen to lie within the basal lamina which splits to enclose them (Fig. 7.8). These cells may be contractile: if so, they must affect capillary diameter.

Sinusoids

Sinusoids are big, baggy capillaries. Instead of flowing down a fine tube of 5–8 μm diameter, the blood flows into an irregular, wide space. The wall is similar to that of a capillary, except that any cross-section is likely to include several endothelial cells. The flow of blood is much slower in a wide sinusoid than in a narrow capillary. Sinusoids often occur in biological filters, where unwanted materials are removed from the blood by macrophages lying on the luminal side of the endothelium. Sinusoids can be fenestrated, or even, as in the spleen and liver, have demonstrable gaps in their walls. Sinusoids are found, apart from these two sites, in the bone marrow and some endocrine glands.

Exchanging vessels and connective tissue

Whether we look at capillaries or sinusoids, a basal lamina lies on the outer side of the tube, beyond which lies the extracellular space of connective tissue. Wherever in the body blood-vessels occur, in muscle, in glands, in the brain, they are carried in a sheath of

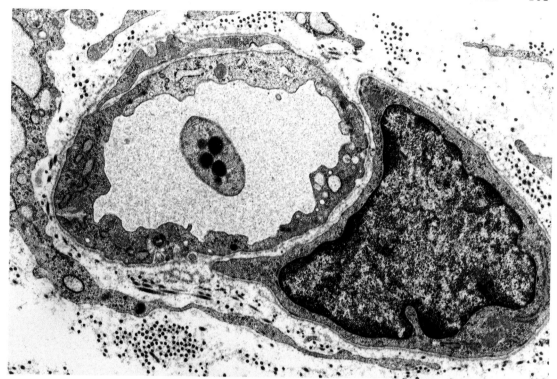

*Fig. 7.8   TEM of capillary with pericyte, from rat uterus. The structure in the capillary lumen is a platelet.
(× 12 000.) (Courtesy, Miss C. Lunam.)*

connective tissue. At first sight, the capillary bed on the alveoli of the lung (Fig. 4.8) would
seem to be in such close contact with the air spaces on either side that there is little room
for connective tissue. But the electron microscope has confirmed that, in places, a basal
lamina surrounds the capillaries, another basal lamina lies beneath the very flattened
epithelium, and the two basal laminae are separated by a space containing collagen; in
other places, a single basal lamina separates the two cell layers (Fig. 7.9). In the
glomerulus of the kidney, the fluid forced out of the "arterial" end of the capillary bed

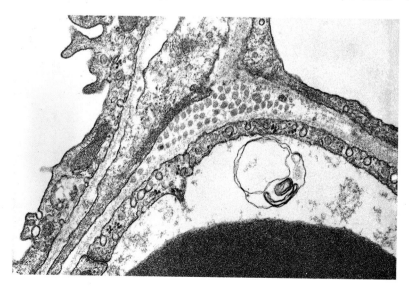

*Fig. 7.9   TEM of epithe-
lial cells lining two alveoli of
the lung (left and above
right), and a capillary lined
by endothelium (below). The
basal lamina underlying each
cell is separate (centre) but
they fuse into a single lamina.
(centre right). (× 40 000.)
(Courtesy, Dr J. Fanning.)*

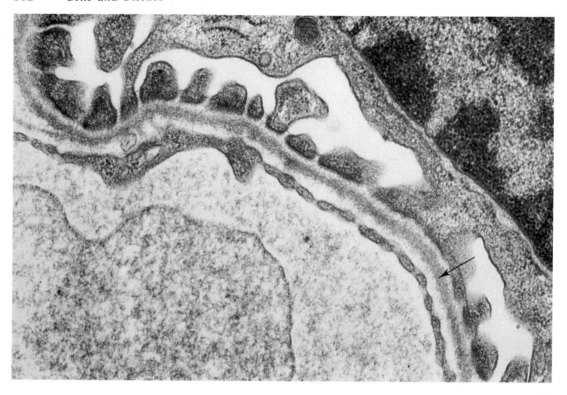

Fig. 7.10   TEM of a glomerulus from the kidney. A capillary (lower left) with a fenestrated endothelium shares a common basal lamina (arrowed) with foot processes of an epithelial cell. Above these processes is the urinary space, with the nucleus of the epithelial cell at top right. (× 38 000.) (Courtesy, Dr S. Bradbury.)

enters the kidney tubule to become the initial filtrate which is converted into the urine. Here, fenestrated capillaries are separated by a single basal lamina from the epithelial cells lining the kidney tubule (Fig. 7.10). These cells have elaborate foot processes resting on the basal lamina, with filtration slits between them. The slits are bridged by a membrane so fine that its very existence has been the subject of debate. Away from the actual surfaces involved in filtration, the shared basal lamina splits into two, separated by connective tissue.

The effects of damage to endothelial cells

Normally, endothelial cells present a smooth, shiny surface to the blood, and the suspended cells circulating past have no tendency to attach to this surface. If, for any reason, the endothelium is damaged, it becomes sticky, for white cells and platelets attach themselves to the blood-vessel wall, and even to each other, implying the local release of a substance producing "stickiness". Injured areas of blood-vessel produce an enzyme, thromboplastin, which causes the conversion of plasma prothrombin into thrombin, an enzyme which, in its turn, catalyses the conversion of the soluble plasma protein, fibrinogen, into insoluble strands of fibrin. The fibrin forms a feltwork in which aggregations of platelets release their granules, and coalesce into a gel-like mass. This fusion of platelets results in the liberation of more thromboplastin and the precipitation of more fibrin contributing to the blood clot.

Many other factors have now been identified as assisting in the formation of clot, or in preventing it in undamaged vessels, and no doubt the long list will continue to grow. However complicated it becomes, clot formation has evolved to seal a hole in the wall of a blood-vessel, and to prevent the circulation of blood through one that is damaged. It is thus of life-saving importance, and those who inherit a defect which prevents blood from clotting, haemophiliacs, have a short life expectancy, sustaining serious damage from bumps and cuts which, in us, would be quite insignificant. Blood that escapes into the tissue also clots. Such clot is ultimately removed by macrophages.

You may wonder what happens if a blow, for instance, results in the closure by clot formation of many small vessels in one area of tissue. How is the blood supply maintained? It seems that an inadequately perfused tissue somehow stimulates the formation of new blood-vessels from surrounding, uninjured vessels. Endothelial cells start to round up and divide, producing a knot of cells at the side of a functioning capillary, and this then grows out as a solid rod of cells. If and when it makes contact with another such rod, both develop a lumen, forming a new capillary. In this way, new vessels grow into the damaged area, while the old, thrombosed ones are broken down and removed by phagocytosis.

Further reading

Askenase, P.W., Graziano, F. and Worms, M. (1979). Basophils and mast cells. Immunobiology of cutaneous basophil reactions. *Monogr. Allergy* **14**, 222–235. Review of basophils' role in the control of local hypersensitivity, and in reactions against ticks and other parasites.

Lewis, S.M. (1974). The constituents of normal blood. *In* "Blood and Its Diseases", Hardisty, R.M. and Weatherall, D.J. (Eds) pp. 3–67. Blackwell Scientific, Oxford.

Vincent, P.C. (1974). Granulocytes and mono-cytes. *In* "Blood and Its Diseases", Hardisty, R.M. and Weatherall, D.J. (Eds) pp. 122–172. Blackwell Scientific, Oxford. Both of these chapters are well written and illustrated: a new edition is due in 1982.

Weller, P.F. and Goetzl, E.J. (1980). The human eosinophil. *Am. J. Pathol.* **100**, 793–820. A comprehensive and up-to-date review.

Zucker, M.B. (1980). The functioning of blood platelets. *Scient. Am.* **242:6**, 70–89. Discusses their roles in haemostasis, atheroma and thrombosis.

Immunity against foreign material

Inevitably cells die and break down in the fluid-filled spaces of connective tissue; inevitably, too, proteins from the blood find their way in small amounts into the extracellular fluid. Yet it is essential that the protein content of the extracellular fluid should be kept to a constant, low level.
Why is this? (Note 8.A)
One mechanism for keeping the protein content low is phagocytosis by macrophages: these digest the protein inside the cell, so that any osmotic effect on the extracellular fluid from splitting large molecules into many small ones is avoided. The other mechanism is the slow collection of fluid and protein in lymphatic capillaries, which ultimately return them into large veins.

In the course of evolution, this pathway for the removal of the body's own proteins from the extracellular spaces has become linked to a system of cells specialized for response against foreign proteins: the latter are very often components of invading micro-organisms or viruses. The cells concerned, lymphocytes, operate in close collaboration with macrophages, and the arrival of foreign proteins at a group of macrophages stimulates a protective reaction, called an immune response, from associated lymphocytes.

The removal of proteins from the extracellular space

The structure of the blindly-ending sacs from which lymphatic capillaries take origin has been described in Chapter 5. The endothelial cells which line them are not sealed to one another by tight junctions, as in a blood capillary, but lie loosely overlapping one another; the basal lamina outside this endothelium also has gaps in it. Fine, collagenous fibres anchor endothelial cells to surrounding structures, so that an increase in extracellular fluid locally separates these cells rather than collapsing the lymphatic sac. Fluid, with dissolved proteins and particulate debris, is thus able to enter the sac freely. On compressing the tissues locally, the endothelial cells come together in their overlapping pattern, preventing the emptying of the sac into the extracellular fluid. Instead, the protein-containing fluid is forced up into a lymphatic capillary.

The transport of lymph

Lymphatic capillaries converge on larger vessels, which are lined by endothelial cells and surrounded by collagen (Fig. 8.1). This structure tells us at once that they cannot propel the fluid up the lumen, or even regulate its flow, as there is no component of the wall that is

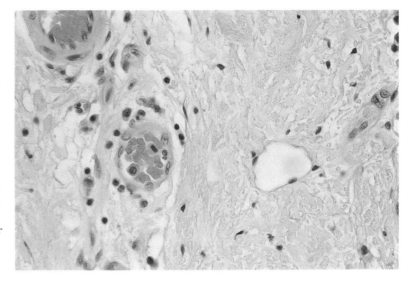

Fig. 8.1 Connective tissue, showing two small blood-vessels (left) and one lymphatic vessel (right of centre) of similar diameter. H & E. (× 385.)

capable of contraction. All the collagen can do is to provide a limit to the distension of the lymphatic vessel. How then is lymph propelled? At many positions in the vessel, flaps of endothelium, each with a delicate core of connective tissue, extend into the lumen to act as valves, allowing the flow of lymph centrally from the periphery, but preventing reverse flow. The actual mechanism for propelling the lymph must be looked for outside the tube. Movement, particularly contraction of surrounding muscle groups, squeezes lymphatics locally, and, given the valves, they can only empty in one direction. Nearer the openings of

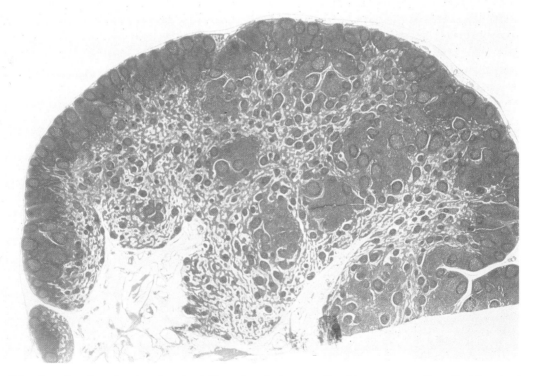

Fig. 8.2 A section through a lymph node. Afferent lymphatics enter the upper, convex surface: the hilum is at bottom centre. Heavily stained aggregations of cells form germinal centres. H & E. (× 14.)

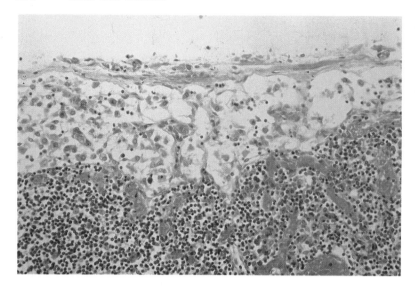

Fig. 8.3 The marginal sinus, lying just below the fibrous capsule of the lymph node (top). H & E. (× 200.)

the large lymphatic vessels into veins, the flow of blood past the opening may entrain lymph, drawing it into the vein. Neither mechanism is likely to produce a brisk flow of lymph, and in fact fluid moves slowly and sporadically up lymphatics.

The filtering of lymph

Somewhere on the line from extracellular space to great vein, lymph passes through a filter which removes cell debris and soluble proteins before the fluid is returned to the blood. These filters are lymph nodes. The design of these filters is remarkably efficient. Lymph nodes are small, bean-shaped structures, with numbers of afferent lymphatics entering the convex surface and a single efferent lymphatic leaving at the concave part, or hilum (Fig. 8.2). The node has a collagenous capsule, resisting distension. Inside this, the afferent lymphatics empty into a large, baggy space, the marginal sinus, from which many irregular sinusoids drain towards the hilum (Fig. 8.3).

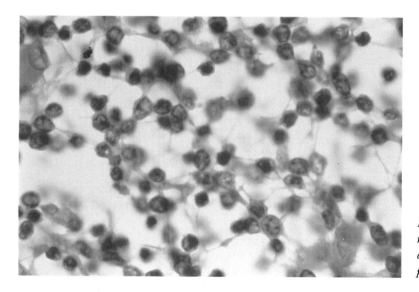

Fig. 8.4 Sinusoids from the medulla of a lymph node, containing free-floating lymphocytes. H & E. (× 1015.)

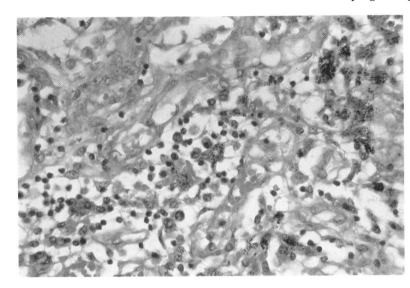

Fig. 8.5 Sinusoids from the cortex of a lymph node from human lung. Macrophages can be seen containing brown, granular material. H & E. (× 405.)

What would you expect to happen to the rate of flow of lymph on entering a large, irregular maze of sinusoids from afferent lymphatics of much smaller diameter? (Note 8.B)
The sinusoids of the lymph node are crossed by strands of collagen running in every direction (Fig. 8.4), and these strands are covered on the side facing the lymph by endothelial cells. These strands further slow the flow rate. As the lymph meanders through this maze, debris and unwanted proteins are removed by a very large number of macrophages, lying on and perhaps between the endothelial cells.

With the light microscope alone, it looks very much as if the macrophages and the endothelial cells are one population, and the older histologists gave the name "reticulo-endothelial cell" to this group of cells, attributing to it the functions of synthesizing the reticular fibres, lining the channel as an endothelial layer and being actively phagocytic. *What are reticular fibres? (Note 8.C)*
With the transmission EM, it is now thought that the endothelial cells and the macrophages are distinct, the endothelial cells synthesizing the fibres. The injection of carbon particles or dyestuffs into the tissues soon results in the accumulation of that material in macrophages in the regional lymph node (Fig. 8.5).

This plan to which the lymph node is designed is found in several other sites in the body where filtering is required, notably the liver and spleen.

The fate of foreign proteins in the tissues

Epithelia are not an absolute barrier to micro-organisms, which frequently get through into the tissues beneath. The tissue fluid which becomes lymph may therefore contain soluble proteins foreign to the body, as well as bacterial cells with foreign proteins on their membranes. The macrophages of the first lymph node encountered will attempt to remove this material, but with very limited success. It seems as if macrophages require a grappling site to hold a protein molecule so that it can be taken into a phagocytic vacuole for enzymes to break it down. Such grappling sites are available for proteins from the host's own body, but a new, completely foreign protein may persist for some time without macrophages appearing able to deal with it. Foreign bacterial cells, too, may resist phagocytosis, or, when phagocytosed, may persist within the cytoplasm of the macrophages.

The immune system responds to the presence of these foreign proteins. The injection of a foreign, soluble protein into an animal which is encountering it for the first time, is followed within 6–10 days by the appearance of new globulin molecules in the blood and tissues which can combine specifically with the foreign protein: these are called antibodies. Since these globulins are either dimers or tetramers, and may attach at more than one site on the foreign protein, large aggregates may form of antibody with the foreign protein, or antigen, and these aggregates may precipitate out of solution. Such antigen-antibody complexes can be readily phagocytosed and digested.

What cell type, apart from macrophages, is believed to be especially active in the phagocytosis of antigen–antibody complexes? (Note 8.D)

Most foreign proteins are antigenic, or capable of stimulating antibody formation, as are some polysaccharides.

If, some months later, the same animal receives a second injection of the same foreign protein, antibodies are produced more rapidly and in far greater quantity: this is the secondary response. The second injection need not be in the same place as the first, or even in the region drained by the same lymph node. Not only have the animal's tissues some "memory" of the first injection, but this information is available all over the body.

A somewhat similar sequence follows the introduction to the body of cell-bound antigens. Some of these stimulate the production of soluble antibodies, in the same way as soluble antigens do. Others, however, produce a cell-mediated immunity, which is effective without being associated with high levels of circulating antibody. This can be most conveniently studied with foreign skin grafts, though the system has clearly evolved to deal with foreign micro-organisms, and perhaps with bizarre cells produced within the body by somatic mutation. About 10 days after a foreign skin graft, the foreign cells die off in great numbers. This is a very specific effect, and host cells just beneath the graft remain healthy. If a second graft from the same donor is made some months later, a secondary response occurs with destruction of the graft in about 5 days: this response is seen anywhere on the body and not just on the site of the first graft. Once again, the first graft is "remembered", and the information is available all over the body. This killing of foreign cells need not be associated with high levels of circulating antibody.

Lymphocytes

Lymphocytes are the cells concerned with these immune responses.

What are lymphocytes like? (Note 8.E)

Apart from their presence in the blood, lymphocytes are found throughout the tissues, and major aggregations of them occur wherever there are groups of macrophages, such as those in lymph nodes. Though there is no easy way for you to distinguish them on your slides, some lymphocytes are specialized for the immune response that produces soluble antibodies — humoral immunity — and others for the response that is cell-bound — cell-mediated immunity. The lymphocytes associated with circulating antibodies are called B-lymphocytes, the lymphocytes responsible for cell-mediated immunity, T-cells. Both are derived from undifferentiated lymphocytes produced in the bone marrow, which, in birds, differentiate into B-cells in an organ called the bursa of Fabricius, and into T-cells in the thymus. The mammalian equivalent of the bursa is not known, though B-cells producing humoral immunity clearly exist. So the circulating blood contains at least 3 subgroups of lymphocytes — undifferentiated, B- and T-cells.

Lymphocytes leave lymph nodes in great numbers, travelling up the efferent lymphatics to enter the blood stream. If one cuts through the main lymphatic channel, the thoracic duct, just before it enters the left subclavian vein, the collected lymph will, within a few hours, contain more lymphocytes than there are in the entire blood volume. This observation led to many hypotheses about the fate of lymphocytes, and it was not until the development of adequate techniques for labelling and tracing cells that the answer emerged.

Suggest a suitable technique for labelling and tracing cells such as lymphocytes. (Note 8.F)

Lymphocytes circulate continually. They leave lymph nodes and the numbers of smaller lymphoid nodules that occur throughout the body to enter the blood. Then, on an apparently random basis, they leave the blood to re-enter lymphoid tissue: this they do by passing through the walls of small venules in lymph nodes and nodules in great numbers, so that their average stay in the blood is only about 2 hours. The vast majority of these circulating lymphocytes are B- and T-cells, each carrying the ability to respond to one antigen, either soluble or cell-bound. The small subgroup of lymphocytes able to respond to one particular antigen constitutes the "memory" described earlier, while its constant circulation between the lymphoid tissues of the body explains the ability to produce a secondary response wherever the antigen may appear.

The immune responses

Let us now look at the immune response which is most clearly understood — the secondary response to soluble antigens — to see how lymphocytes function. We will go on from there into cell-mediated immunity and primary responses.

The secondary response to soluble antigens

We start with a small subpopulation of B-cells, circulating between the body's lymphoid tissues, carrying the "memory" of one particular antigen. The arrival of a second dose of that antigen in the tissues results in a large number of macrophages in the local lymph node or nodule being challenged with a foreign protein. Somehow (and we do not understand the details of the collaboration of macrophages and lymphocytes), B-cells of this subpopulation immediately adjacent to some of these macrophages are "notified" of the arrival of the antigen. Two events follow: first, the small subpopulation grows rapidly in size by mitosis; then, the cells so produced start to synthesize and release antibody molecules. Each of these two steps, which are not separate but overlapping in time, results in a change in appearance of the lymphocytes concerned.

For a cell as inactive as a "memory" lymphocyte to divide, the genes for cell division must be switched on and a whole series of syntheses must start, producing new DNA and chromosomal proteins, new proteins for the mitotic spindle, sufficient cytoplasm and organelles for the two daughter cells and so on.

What changes would you expect in the cell's structure and appearance as it prepares to divide? (Note 8.G)

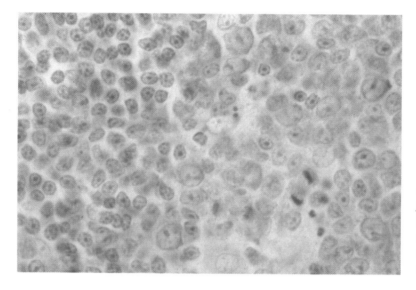

Fig. 8.6 The edge of a germinal centre, showing large lymphocytes, some of which are dividing, to the right, and small lymphocytes to the left. PMG. (× 1015.)

These changes are known as transformation of the lymphocyte, and they produce a cell known to the older histologists as a medium or large lymphocyte (Fig. 8.6).

When the daughter cells of transformed lymphocytes start to synthesize antibody, they again change their appearance.

What structure and appearance would you expect in a cell that synthesizes a single protein in large amounts for continuous secretion? (Note 8.H)

This appearance is so different from that of the original "memory" cells that it was given a different name — the plasma cell (Fig. 8.7). So the three names, small and large lymphocyte and plasma cell, can all refer to the same cell type at different periods in its existence, and intermediate appearances exist between the various stages. The conversion of a small population of "memory" B-cells into a very much larger one of plasma cells takes several days, but produces the high levels of circulating antibody seen from day 5 onwards. As the antigen is removed, the stimulus for antibody production goes too, and plasma cells appear to revert to "memory" cells again, though some maintain that cell division in the transformed lymphocytes produces two distinct types of daughter cell — "memory" and

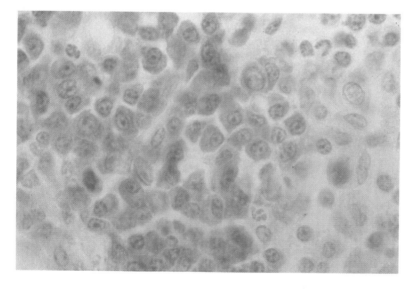

Fig. 8.7 Plasma cells within a lymph node. PMG. (× 1015.)

plasma cell — rather than the latter turning into the former. Plasma cells are usually stationary, and can be seen in large numbers around an infected site and in the local lymphoid tissue.

The secondary response to cell-bound antigens

This has many similarities to the humoral secondary response. In fact, some cell-bound antigens provoke a typical humoral response, with high levels of circulating antibody — these are seen, for instance, in haemolytic anaemias. Other cell-bound antigens, however, trigger a different chain of events. The arrival of cells carrying the antigen challenges macrophages which "notify" nearby "memory" T-cells of the subpopulation responsible for immunity against that particular antigen. This message is a stimulus, transforming the lymphocytes into dividing cells and producing a great expansion of the subpopulation. The final effector cells, however, are very different from the plasma cell. They look just like "memory" cells, though they appear to have on their cell membrane antibody molecules capable of combining with the antigen on the membrane of the foreign cell. Effector T-cells (sometimes given the dramatic names of cytotoxic or killer cells) seem to attach to the surface of the foreign cells and kill them by disrupting the membrane.

The primary immune response

Given an appropriate subpopulation of "memory" cells, it is not too difficult to follow through the humoral and cell-mediated immune responses. The early stages of the primary response are more difficult, and there are various explanations available.

One hypothesis supposes the existence of populations of B- and T-cells which have not yet become committed to any particular antigen. In this case, the arrival at a macrophage of a foreign protein which is new to the animal is followed by information of some sort passing to one of these uncommitted lymphocytes, which develops the ability to respond to that particular antigen. The cell then grows, transforms and prepares for mitosis, producing a clone of similar cells, a subpopulation ready for further expansion and response: events from here on would be similar to those seen in the secondary response. The mechanism by which an uncommitted B- or T-cell would acquire the ability to deal with a new antigen is quite unknown.

An alternative hypothesis supposes that in late foetal and early post-natal life, committed B- and T-cells are produced for every conceivable antigen, and that these survive indefinitely until stimulated by the arrival of their particular antigen. In this model, the much slower primary response is due to the very small size of the circulating population of cells committed to dealing with that antigen: this may delay the initial "notification" of a committed cell, and would clearly require more cycles of cell division to build up an effective subpopulation to respond to the antigen.

Lymphocytes and lymphoid tissue

Immunology is a complex and rapidly growing science. Enough is already known, however, to allow us to recognize histologically some of the stages of the various immune responses and to understand the organization of lymphoid tissue.

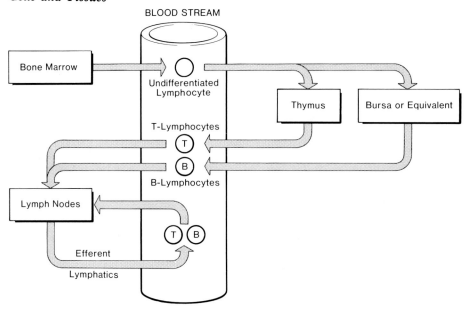

Fig. 8.8   The patterns of circulation of lymphocytes.

The life histories of lymphocytes

When we see a small lymphocyte in a smear of blood cells (Fig. 7.5), it can be one of three different groups of lymphocyte, and histological observation does not help us to distinguish the one from the other.

Lymphocytes are produced post-natally in the bone marrow (Fig. 8.8), and they enter the blood stream undifferentiated. From the blood, some enter the thymus, where they differentiate into T-cells, capable of cell-mediated immune responses. There is now considerable evidence that lymphocytes in the thymus are by and large incapable of cell-mediated responses, but lymphocytes leaving the thymus have that ability.

If the thymus is the site of differentiation of T-cells, what might be the result of removing the thymus from a newborn animal? (Note 8.1)

Differentiated T-cells leave the thymus for the blood stream, and enter the endless circulation of peripheral lymphocytes between lymphoid tissues, coming into close association with macrophages in each node or nodule.

Other undifferentiated lymphocytes from the bone marrow pass through the mammalian equivalent of the bursa of Fabricius to emerge as B-cells: it has been suggested that some of the lymphoid tissue in the wall of the gastro-intestinal tract may be this site of differentiation. Once differentiated into B-cells, these lymphocytes also enter the circulation routes between lymphoid tissues.

The organization of lymphoid tissue

The bone marrow and the thymus are specialized tissues for, amongst other things, the production of undifferentiated lymphocytes and the conversion of undifferentiated lymphocytes into T-cells. They do not have the filtering functions of other lymphoid tissues, so they lack the basic structure common to all the others.

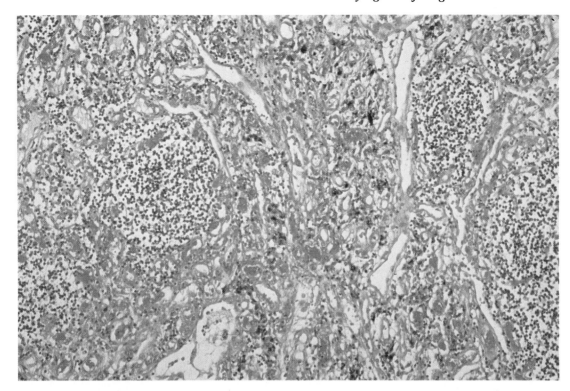

Fig. 8.9   The cortex of a lymph node from human lung. Sinusoids can be recognized from the brown, granular material in macrophages. Clusters of lymphocytes lie between the sinusoids. H & E. (× 145.)

In lymphoid filters, structural modifications ensure a very slow flow rate.
How is this slow flow rate achieved? (Note 8.J)
The fluid filtered may be lymph, which is essentially extracellular fluid from tissues in the region drained by the lymph node or lymphoid nodule, or blood, as in the spleen. As the fluid drifts slowly past, large numbers of macrophages lining the sinusoids actively phagocytose and digest soluble proteins and damaged cells. Closely associated with these macrophages in the loose connective tissue between the sinusoids are very large numbers of lymphocytes, most of which are the typical, small lymphocyte (Fig. 8.9).

Lymphoid tissue usually contains evidence of immune responses, in addition to its necessary populations of lymphocytes and macrophages. Both humoral and cell-mediated responses, primary and secondary, contain a common step in which transformed, large lymphocytes divide repeatedly to produce a very much expanded population. Clusters of dividing large lymphocytes are often seen (Fig. 8.6), and produce an appearance known as a germinal centre. In the lymph nodes, such germinal centres are often seen in the outer rim, or cortex, of the node.
Would you expect to see germinal centres in animals born and reared in a germ-free environment? (Note 8.K)
In addition to this evidence of expansion of subpopulations of lymphocytes, lymphoid tissues normally contain many plasma cells involved in antibody synthesis (Fig. 8.7).

Whereas most of the lymphocytes in lymphoid tissue lie between the sinusoids, many enter the fluid within the sinusoids, whether it be lymph or blood, to circulate to other lymphoid tissues. So, in lymph nodes, the marginal sinus contains few lymphocytes, but the sinusoids in the medulla, as they converge on the efferent lymphatic, are full of them.

Apart from the large lymph nodes, smaller nodules of lymphoid tissue lie beneath most epithelia, and are particularly common and large beneath epithelia with a high probability of penetration by micro-organisms.

If micro-organisms do get into the tissues in any number, the infected site may have many macrophages active around it, and many lymphocytes, particularly T-cells may collect there. This infiltration of tissues with lymphocytes (Fig. 15.8) lacks the organization typical of lymphoid tissue, and should not be confused with it.

If a new antigen is introduced into the tissues, whether it is soluble or cell-bound, the earliest and greatest immune response takes place in the lymph nodes draining lymph from the tissues concerned. At an early stage of the response, transfer of these lymph nodes (or even cells from them) to a host animal genetically similar enough to accept the graft will result in the immune response continuing in the host, even though it has never encountered the antigen. Transfer of other lymph nodes or cells at this early stage in the response does not transfer immunity. If the new antigen is introduced directly into the blood stream, the earliest and greatest response is in the spleen. After these initial stages of the primary response, immunity can be transferred by grafting any lymph node, or the spleen, or cells from them or from the thoracic duct.

The immune system throughout life

As the very young embryo develops and differentiation of many cell types begins for the first time, "new" proteins will appear in the membranes of cells and in the extracellular fluid. It would clearly be disastrous if these were to stimulate an immune response. It seems as if macrophages can develop the molecular machinery to cope with new antigens during foetal life, without the need to call on lymphocytes for help. At about the time of birth, macrophages start to "inform" the immune system when challenged with a new antigen. Interestingly enough, foreign cells injected into a mammalian foetus provoke no immune response at the time, and many continue to live quite happily in the host throughout its lifetime. This happens in some species when unlike twins share the uterus during pregnancy: in such a case, the twins may accept skin grafts from each other later in life without rejection. The newborn animal accepts as "self" any macromolecule its macrophages have had to deal with up to that time, and only recognizes as "non-self" new proteins, encountered for the first time after the immune system begins to function.

It is statistically very likely that the rate of meeting new antigens will be higher in the young animal than in one that has lived for some time. Lymphoid tissue is relatively larger and contains more germinal centres in young animals; lymphoid tissue regresses and contains fewer germinal centres in older animals.

There are a few very specialized regions of the body from which the proteins do not normally reach the lymphoid tissues. The eye is one example. The proteins of the lens of the eye, if removed and injected into the connective tissues beneath the skin of the same animal, will provoke an immune response. This response to one's own tissues is called auto-immunity. It sometimes happens that common tissue proteins can become antigenic, probably as a result of combination with toxins produced by invading micro-organisms, with drugs, or as a result of viral infections: the complex of tissue protein and foreign reactive group is treated as a new antigen, but some of the antibodies produced react against the tissue protein. Auto-immune diseases like this are difficult to treat.

Note that immunity involves antigenic material in the extracellular fluid or exposed on the surfaces of cells. The inside of the cell's cytoplasm cannot distinguish foreign material. Cells can be fused together experimentally by treatment with Sendai virus under specific conditions, and it is quite possible to fuse a chick cell with a human cell and find both nuclei living happily within the same cell membrane. It is this safe, intracellular position that viruses occupy while they are replicating: they are vulnerable to immune attack only when they appear in the extracellular spaces, unless a tell-tale viral antigen is incorporated in the cell membrane, making the cell a target for destruction.

Further reading

Cooper, M.D. and Lawton, A.R. (1974). The development of the immune system. *Scient. Am.* **231:5**, 58–72. A clear account of humoral and cell-mediated immunity.

Ford, W.L. (1974). The formation and function of lymphocytes. *In* "Blood and Its Diseases", Hardisty, R.M. and Weatherall, D.J. (Eds) pp. 173–222. Blackwell Scientific, Oxford. Fuller survey than Cooper and Lawton's, but equally readable.

Goldschneider, I. (1980). Early stages of lymphocyte development. *Curr. Topics Dev. Biol.* **14:2**, 33–57. Summarizes T-lymphocyte differentiation up to the point of the antigenic stimulus. Covers more recent work than previous two references, but more specialized, jargon-filled vocabulary.

Makinodan, T. and Kay, M.M.B. (1980). Age influence on the immune system. *Adv. Immunol.* **29** 287–330. Review of recent work on age changes and immunity.

Raff, M.C. (1976). Cell-surface immunity. *Scient. Am.* **234:5**, 30–39. Antigen-antibody reaction on the cell membrane.

Rosenthal, A.S. (1980). Regulation of the immune response — role of the macrophage. *New Engl.J.Med.* **303**, 1153–1156. Brief review of receipt of antigens by macrophages, and their interactions with lymphocytes to initiate the immune response.

Contraction and muscle

Up to now, in describing cells, their shapes and the various phases of their lives, we have taken many things for granted. How is cell shape maintained? How does a cell change its shape? How do cells move? It is now time to look at these mechanisms in detail, both in those cells which are specialized for moving the body as a whole and in the individual cells which move about within the body. For it has become clear over the past few years that the same mechanisms produce contraction in the cytoplasm of the macrophage and in the muscles of the leg. The older histologists placed muscle cells in a separate group in their classifications of cells and tissues. It now seems that the ability of cytoplasm to contract is common to nearly all cells, and that the mechanism of contraction is essentially similar. Muscle cells are specialized only in investing more of their resources in the common contractile mechanism.

It is also clear now that the mechanism that produces contraction can be modified in a stationary cell to resist deforming forces. Cells that move at one time and take up a fixed position at another use the same basic mechanism in both activities. In this chapter, then, we will look at contractility, at how it is brought about and at how the system changes in a stationary cell, rather than confining ourselves to a discussion of the specialized tissue, muscle.

The major proteins of contraction

Many of the mechanisms for cell movement, and for the maintenance of cell shape when movement ceases, centre around three families of proteins. Within each family, there are different molecules with slightly different functions, but the similarities are for our purpose much more important than the differences, and I shall talk of actin, myosin and alpha-actinin from now on as if each were only one protein.

Actin is a filamentous protein, a polymer built up out of repeating monomers; the change from monomers to polymer is reversible, and seems to happen with surprising rapidity in the cell. The filaments of actin can be seen with the TEM, and have been called microfilaments or tonofilaments: they have a diameter of about 6 nm. Actin filaments are not reversible like lengths of string, but have a definite polarity. One end of the filament attaches to another protein, alpha-actinin, while the filament itself can only interact with the active head of the myosin molecule in one direction, as we shall see. It is the commonest of the three proteins, forming about 15% of the total protein in actively mobile cells, such as *Amoeba*, and about 2% of the protein in fixed cells such as the epithelial cells of the liver.

Alpha-actinin is a protein that attaches itself to the basal end of actin filaments. It is capable of cross-linking to other molecules of alpha-actinin, forming plates or discs from which numbers of actin filaments arise. It can attach to the protein layer on the inner or cytoplasmic side of the cell membrane.

Myosin, the third of the proteins involved in contraction, is again filamentous. Unlike actin, it is double-headed; each end of the filament carries little barbs, called heavy meromyosin, so that it can be thought of as an arrow with a head at each end. The filaments of myosin vary in diameter in different cells: in non-muscle cells, their diameter is about 6 nm, while in muscle it may be as high as 15 nm.

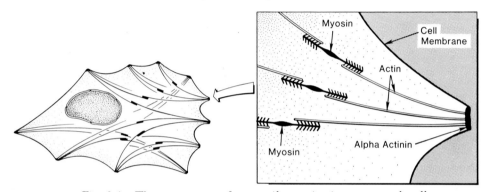

Fig. 9.1   The arrangement of contractile proteins in a non-muscle cell.

The mechanism of contraction (Fig. 9.1) involves actin filaments from two different anchoring sites coming into contact with a myosin filament. In the presence of ionic calcium, myosin in contact with actin splits the terminal phosphate group from ATP, and uses the energy liberated to move up the actin filament towards the alpha-actinin. As the myosin does this at both ends, climbing simultaneously up two actin filaments, it effectively pulls closer together their two anchoring sites, where they attach to alpha-actinin. In the absence of ionic calcium, the myosin disengages from the actin, allowing the two anchoring sites to move apart. In the absence of ATP, the ratchet-like movement of myosin along the actin filaments cannot take place.

Movement in non-muscle cells

If living cells are watched at high magnification, many cell types show continued, rapid movements of cytoplasmic organelles, movements which seem altogether random. This Brownian movement occurs in any liquid, including cytoplasm. The surface membrane of the cell, however, often displays movements which are much slower, and which seem much more purposeful.

How can living cells be observed in the light microscope? Can they be watched by electron microscopy? (Note 9.A)

Patterns of movement

The cell type most frequently studied is the fibroblast, which in tissue culture often moves over the substrate. The cell is usually elongated, with a rather short advancing edge (Fig. 5.3). This advancing edge extends projections called ruffles, which consist of thin, watery cytoplasm between two layers of membrane, forming a wavy, curtain-like extension of the cell. A ruffle may dip down and make contact with the substrate on which the cell sits, or it may move back on to the upper surface of the cell and disappear. If it establishes contact with the substrate, it seems to act as an anchoring site, and the cell pulls itself forwards towards this attachment, letting go at attachment points nearer the rear of the cell.

This pattern of movement consists of two distinct phases. In the first, the ruffle is extended ahead of the cell body; in the second, the cell is pulled forward towards the point of attachment to the substrate. Both are achieved by the same mechanism of contraction. Recent studies using fluorescent antibodies against actin, myosin and alpha-actinin have shown that the two phases are associated with different patterns of organization of these molecules within the cell. Extension of the ruffled membrane is associated with an apparently random arrangement of actin filaments to form a very fine meshwork in the cytoplasm. One can visualize myosin molecules pulling together actin filaments in such a meshwork, squeezing the watery cytoplasm out into a ruffle or cell process. Once the ruffle attaches to the surface of the glass slide, the attachment point seems to attract alpha-actinin, which forms a disc at the membrane with actin filaments radiating back from it into the cytoplasm. Other sites on the cell membrane, including the other points of contact of cell with substrate, have similar discs with radiating filaments. Myosin molecules, pulling together filaments from the front attachment point and those from elsewhere in the cell, tend to move the cell towards the front attachment, provided it is tightly enough fixed to the substrate.

Given different patterns of organization of the actin filaments, then, the same contractile mechanism can extend and retract processes, or can alternate the two so that the cell creeps slowly forward over the substrate.

The varying patterns of actin

So actin filaments can exist in two major forms: as a fine meshwork in the cytoplasm or as bundles of parallel filaments, usually attached to the cell membrane by alpha-actinin at one end. These two forms are interchangeable, and, in addition, the polymeric actin of the microfilaments can be broked down into soluble monomers, with which it is in equilibrium. In a free-floating, spherical cell which is continually putting out cytoplasmic processes, such as a free macrophage, most of the actin is present as the filamentous polymer, and these filaments form a dense meshwork through the cytoplasm. A cell that is moving over a surface has a leading edge containing a fine meshwork of actin filaments, and, in addition, a number of attachment sites from which bundles of actin radiate out into the cytoplasm. At the advancing edge, the meshwork pattern is converted into the bundles as new attachment points are formed, while at the rear of the cell attachment points disappear, with the conversion of bundles into meshwork again, and probably conversion of filamentous actin into the monomeric state. In cells that have ceased moving and have formed firm contacts with surrounding structures, nearly all the demonstrable actin forms bundles radiating into the cytoplasm from attachment points on the cell membrane.

Name two specializations of the membrane which have this pattern. (Note 9.B)

The rapidity with which filamentous actin can be converted into the monomeric form is well illustrated in mitosis. After the separation of the two sets of chromosomes, a furrow appears in the cytoplasm between the newly formed nuclei; this furrow deepens, through the stage in which the cells are joined only by a thin rod of cytoplasm, until complete separation of the cells occurs. It has been shown that the furrow is associated with a complete ring of actin filaments, lying just below the cell membrane. Myosin molecules in this ring gradually draw it closed, yet the bundle of actin filaments remains the same thickness while the diameter of the ring diminishes. This would clearly be impossible if all the actin filaments initially present remained there throughout the constriction process — the bundles of filaments would have to thicken as the diameter of the ring shrinks. Conversion of actin from the filamentous polymer to the soluble monomer must be proceeding in step with constriction.

A moving cell becomes stationary

As a moving cell becomes stationary, actin filaments change from the fine meshwork to bundles of filaments attached to the cell membrane. In doing so they pass through a very interesting intermediate state, producing a regular, three-dimensional network of filament bundles resembling a geodesic dome over and around the nucleus. Each of the triangular "panels" of this dome has demonstrable alpha-actinin at its corners, and these corners seem to act as organizing centres for the further aggregation of filaments from the fine meshwork. This perinuclear dome develops thick bundles of filaments radiating out to the cell periphery. Finally, as the cell becomes attached, the dome disappears, leaving the filament bundles. Alpha-actinin is present not only at the attachment points of these bundles to the membrane, but also at sites along the bundle.

Cells that are not only stationary but required to resist separation from their neighbours have so many bundles of actin filaments in their cytoplasm that these can be detected by light microscopy in good preparations. The cells of the stratum spinosum of epidermis and of the oesophageal epithelium come into this class. Each is held to its neighbours by hundreds of desmosomes, and the actin bundles radiating into the cytoplasm from these attachment sites give the cell a coarsely fibrillar appearance. It is not at present clear how the actin filaments from one desmosome are attached to those from another. It is tempting to suggest that they are linked by some myosin-like molecule that lacks the ability to move along the actin, causing contraction, but is strong enough to resist forces that would pull the filaments apart.

Muscle cells

Some cells are so highly specialized for contraction and attachment that their cytoplasm is largely filled with actin filaments, and with the associated proteins, myosin and alpha-actinin. These are muscle cells, which are classified into three groups — smooth, striated and cardiac.

Each type of muscle was classified initially on the basis of its appearance in histological preparations. These different patterns of organization produce differences in function, as we would expect. We now realize that each type of muscle can be further subdivided. The similarities between the various subclasses of striated muscle fibre are, however, sufficiently great for us to be able to ignore the differences in this book. We will therefore look at smooth, striated and cardiac muscle in turn, as common patterns of organization of contractile tissue.

Smooth muscle

Smooth muscle is structurally the simplest type of muscle. It consists of single cells, often arranged in bundles or sheets which contract together. It is capable of powerful and sustained contraction, and is economical in terms of energy expended. It is widely distributed in the body, particularly in internal organs and for functions concerned with the internal environment; it produces and regulates flow down many of the hollow, tubular structures within the body. Intermediate forms exist between the smooth muscle cell and the fibroblast, particularly in pathological conditions.

The organization of smooth muscle

Smooth muscle cells can be separated out from one another: they are then seen to be long, thin cells tapering at each end, and with a central nucleus. The nucleus is sausage-shaped, with relatively little condensed chromatin and one or two nucleoli, suggesting the continuing synthesis of a range of proteins.

With the light microscope, the cytoplasm appears fairly homogeneous, except for a hint of filaments running the length of the cell (Fig. 9.2). With the TEM, the central nucleus can be seen with a cap of cytoplasm on either end containing many mitochondria, short lengths of RER and SER, Golgi apparatus and free ribosomes. The rest of the cytoplasm is occupied by fine filaments, running longitudinally (Fig. 9.3). That actin and myosin are present in smooth muscle cells is undeniable, but their precise recognition by electron microscopy has caused some debate. It may be that, as in non-muscle cells, the filaments of myosin are similar in diameter to those of actin. On the other hand, some electron

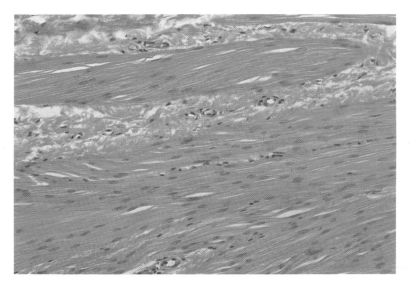

Fig. 9.2 Smooth muscle fibres in the wall of the human intestine. H & E. (× 200.)

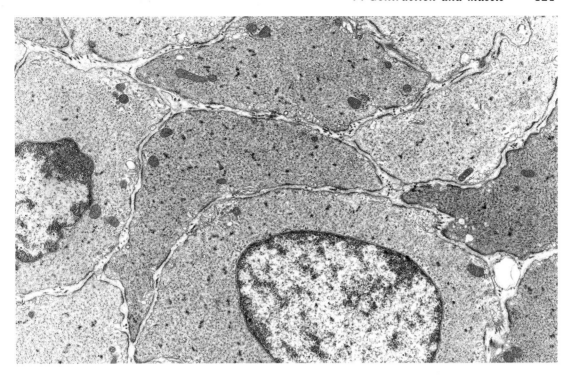

Fig. 9.3   TEM of smooth muscle fibres: in this transverse section, the cytoplasm is largely filled with filaments of actin and myosin. (× 9000.) (Courtesy, Dr B. Gannon.)

microscopists claim to have demonstrated two classes of filament, a thick and a thin, in some smooth muscle cells.

There are many discs on the inner side of the cell membrane, to which filaments are attached (Fig. 9.4): these are alpha-actinin, which also appears at points in the cytoplasm along the bundles of actin filaments. In cells fixed during contraction, dimples appear in the membrane at attachment discs.

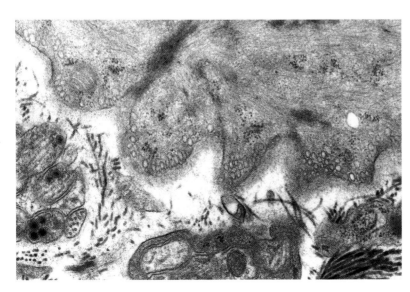

Fig. 9.4   TEM of part of a smooth muscle cell (above). Processes of autonomic nerves lie amongst bundles of collagen (below). Note dense attachment sites on the muscle membrane, with actin filaments radiating from them. (× 32500.) (Courtesy, Dr B. Gannon.)

Contraction, then, involves the ratchet-like movement of myosin filaments along the filaments of actin; since myosin is double-headed, this movement occurs simultaneously along two actin filaments attached to opposite ends of the cell. Contraction shortens the cell, which inevitably gets fatter at the same time.

What two factors are necessary for this reaction of myosin with actin to occur? (Note 9.C)

Smooth muscle cells usually occur in bundles or sheets, with fine layers of connective tissue between the bundles. Even within the bundles, GAGs and fine reticular fibres lie between the individual cells (Fig. 9.3). The attachment discs of alpha-actinin often lie opposite one another on adjacent cells, but there are no typical desmosomes.

What does a desmosome look like, and how might it differ from these adjacent attachment discs? (Note 9.D)

At places, adjacent smooth muscle cells show the membrane specializations associated with gap junctions.

What are these, seen with transmission electron microscopy of sectioned material, and by freeze fracture? (Note 9.E)

The functional characteristics of smooth muscle

Smooth muscle cells are powerful and economical. They contract rather slowly, building up to a maximal shortening over the course of about 60 seconds, but they can hold this contraction for many minutes before fatiguing. Their relaxation is also rather slow. These characteristics make them ideal for many activities in the interior of the body. Smooth muscle regulates the flow of blood down blood-vessels and propels materials down tubes as varied as the intestine, the urinary tract and the reproductive tracts. In all these sites, the economy of smooth muscle in terms of the conversion of chemical energy to contractile power is valuable, and the relatively slow contraction and relaxation pose no particular problem.

Smooth muscle in the walls of many viscera appears to have a spontaneous, rhythmic contraction, which can also be seen in tissue culture. When the muscle is organized in bundles or sheets a wave of contraction spreads across the mass of cells periodically.

What structures are present on the surface of smooth muscle cells which might account for this spread of contraction? (Note 9.F)

In addition to this built-in rhythmic contraction, stimulation of the muscle cells (which can be done through the nerves supplying them, hormonally, directly by electrical means, or even by stretching them) can result in a tonic, sustained contraction.

The recognition of smooth muscle

It is important to be able to recognize smooth muscle fibres confidently in sectioned material. The oval nuclei and the rather featureless cytoplasm, which stains red with eosin, lead many students to confuse smooth muscle with collagenous connective tissue, which stains similarly and contains the elongated, ovoid nuclei of fibroblasts. With care, the bundles of collagen running through connective tissue can be distinguished from the much blander-looking cytoplasm of smooth muscle, and the nuclei of smooth muscle are more regular and usually plumper than those of fibroblasts.

What staining techniques do you know which might help to distinguish smooth muscle from connective tissue? (Note 9.G)

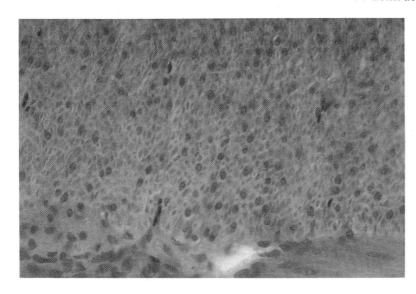

Fig. 9.5 Transverse section of a bundle of smooth muscle fibres from human intestine. H & E. (× 405.)

Within a bundle of smooth muscle cells, the cells are staggered relative to each other, so that the nuclei lie at different levels, and any transverse section through the bundle will show cell profiles varying from the maximum diameter at the centre of the long cell to the thin, tapered end (Fig. 9.5). Nuclei will only be found in the larger profiles, and will be centrally situated.

Striated muscle

This type of muscle is highly specialized to produce very rapid contraction, which is precisely controlled by peripheral nerves. Striated muscle fibres arise from the fusion of many myoblast cells: they are large, multinucleate cylinders rather than single cells. The nuclei, which may number hundreds in large fibres, are not centrally placed in the fibre, but lie just below the surface membrane, which is called the sarcolemma (Fig. 9.6). Within the fibre, longitudinally running myofilaments can often be distinguished, each myofilament made up of repeating structural units called sarcomeres. Since the sarcomeres of all the myofilaments in a single fibre are in register, the result is a regular cross-striation of the fibre, clearly visible with the light microscope (Fig. 9.6).

The structure of a single sarcomere

The organization of actin, alpha-actinin and myosin in a single sarcomere is illustrated in Figs 9.7 and 9.9. At each end of the sarcomere, the Z-disc runs transversely across the myofilament: this disc consists of the protein, alpha-actinin. From each side of the Z-disc filaments of actin extend: if we look at a single sarcomere, these filaments point into the sarcomere in a regular, parallel array from each end. They do not quite meet in the middle of the sarcomere. Lying between the parallel filaments of actin, which are about 7 nm in diameter, are thicker filaments, 10–15 nm diameter, confined to the centre of the sarcomere and never reaching as far as the Z-discs — these are myosin filaments. As in smooth muscle and non-muscle cells, the myosin is double-headed, with little barbs of

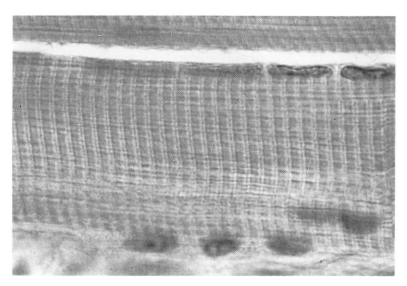

Fig. 9.6 Longitudinal section of part of a fibre of striated muscle. H & E. (× 960.)

heavy meromyosin pointing towards the Z-discs and engaging with sites on the actin filaments. In the presence of Ca^{2+}, the heavy meromyosin barbs split the terminal phosphate from ATP, utilizing the energy to move up the actin filaments towards the alpha-actinin of the Z-discs. Since this reaction happens simultaneously at both ends of each myosin filament, the two Z-discs at the opposite ends of the sarcomere are drawn closer together. In full contraction the sarcomere shortens by about 30% of its resting length.

How are the striations that are visible with the light microscope linked to the molecular structure of the single sarcomere? Figure 9.6 shows alternating dark and light bands along the fibre, called A-bands and I-bands by the early microscopists. In the centre of each light band is a darker line, the Z-line or Z-disc. The single sarcomere lies between two Z-bands, although we must remember that, with the light microscope, we are not looking at a single myofilament but many of them in register within the fibre. The dark A-band in the centre of the sarcomere corresponds to the myosin filaments, and the light I-band with the Z-line in its centre is occupied by the thinner actin filaments. But we know that actin filaments do not end at the level where myosin filaments begin, but penetrate between the myosin filaments for a short distance, forming attachments with myosin and permitting the ratchet-like movement of the heads of the myosin filaments up them. In good preparations, the darker A-band is seen not to be uniform. A lighter central portion can be distinguished, called the H-band. This is the central part of the sarcomere, where myosin filaments only are found: the darker, outer parts of the A-band have filaments of actin lying between the myosin filaments. To complicate things still further, the central part of the H-band, the very centre of the sarcomere, is marked by a denser line, the M-band, which is the site of attachment of the myosin filaments to each other.

Much of the structure of the striated muscle fibre was described long before it was understood. It is an indication of the extraordinary regularity with which the fibre is organized that these striations visible with the light microscope can be described in terms of the protein molecules involved in contraction. The pattern sounds complicated, but is

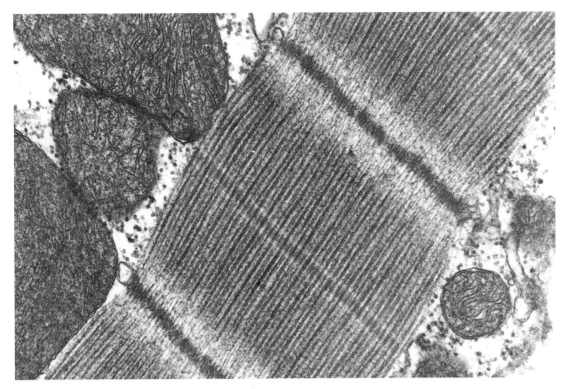

Fig. 9.7   TEM of a single sarcomere from cardiac muscle, fixed in contraction. Large mitochondria lie on either side. (× 58 000.) (Courtesy, Dr B. Gannon.)

really very simple if one remembers the organization of the three contractile proteins that is necessary to produce shortening. The anchoring protein, alpha-actinin, has filaments of actin extending from it: actin filaments from two anchoring sites are pulled closer together by a double-headed molecule of myosin, moving simultaneously up each actin filament towards its anchoring site.

Familiarity with the structure of the sarcomere (Fig. 9.7) allows us to predict the appearance we would expect if a striated muscle fibre is sectioned transversely at different levels, and viewed by the TEM (Fig. 9.9). A section at the level of the I-band, for instance, will show a regular pattern of thin actin filaments; a section through the H-band will show a regular pattern of thick, myosin filaments.

What would you expect to see in a section through the outer, darker segment of the A-band? (Note 9.H)
Similarly, a knowledge of the structure of a sarcomere allows us to predict how the bands will alter in length on full contraction (Fig. 9.8).
What changes in banding will occur on maximal shortening of the muscle fibre? (Note 9.I).

The sarcoplasmic reticulum

We have described the structure of the sarcomere, but we also need mitochondria and a mechanism to make Ca^{2+} available within the fibre if myosin is to react with actin to

Fig. 9.8   TEMs of human striated muscle. (a, × 16500: b, × 10500.) (Courtesy, Dr M. Haynes.)
(a) Fixed in relaxation.

(b) Fixed in full contraction.

produce contraction. Mitochondria occur in linear arrays, lying between the myofilaments (Fig. 9.8). The control of Ca^{2+} concentration is more complicated.

Striated muscle fibres possess an elaborate system of tubules, the sarcoplasmic reticulum, a sort of SER, running longitudinally around each myofilament. Sections of fibres, viewed by the TEM, show a second set of membrane-bound tubules running transversely across the fibre in the cytoplasm surrounding the myofilaments. These T-tubules open on to the surface of the fibre: they are, in fact, invaginations of the surface membrane.

What is the name for the surface membrane of a striated muscle fibre? (Note 9.J)

At regular sites through the fibre, T-tubules come very close to enlarged portions of the sarcoplasmic reticulum, producing an appearance known as the triad. These structures are difficult to visualize, but are presented three-dimensionally in Fig. 9.9.

The sarcoplasmic reticulum pumps calcium ions into its lumen from the surrounding cytoplasm. Stimulation of a muscle fibre to contract produces a leakiness of the surface membrane, resulting in an increased flow of ions across it and loss of the electrical charge

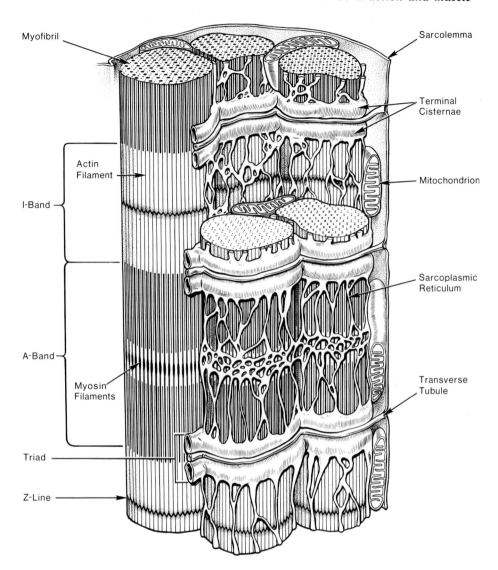

Fig. 9.9   The structure of striated muscle, relating the arrangement of contractile proteins to the cross-striations.

which normally exists across the membrane. The T-tubules are continuous with the surface sarcolemma, and carry the same ion fluxes deep into the fibre. Where the T-tubules come into very close contact with the sarcoplasmic reticulum, the latter also depolarizes, liberating its Ca^{2+} ions into the cytoplasm. Since T-tubules occur frequently and regularly throughout the fibre, depolarization of the surface membrane results in the liberation of Ca^{2+} simultaneously right through the fibre, and simultaneous contraction of all the sarcomeres.

The functional characteristics of striated muscle

The elaborate, repeating organization of contractile proteins has evolved to allow the muscle fibre to contract rapidly, achieving full shortening within a remarkably short time. Such a fibre contracts and relaxes more rapidly than smooth muscle; it also fatigues faster.

Each fibre is surrounded by a fine sheath of connective tissue: there are none of the cell-to-cell communications seen in smooth muscle. Stimulation of a single striated muscle fibre, then, produces a powerful and rapid contraction, which does not spread to neighbouring fibres.

Striated muscle is mainly used for movements of the body relative to the external environment — movements that often need to be powerful, rapid and precisely controlled. The beautiful organization of sarcomeres and tubule systems makes possible the simultaneous contraction of the whole fibre: the size of the largest fibres of skeletal muscle is so great that this abrupt contraction exerts a considerable force. The insulation of each fibre from its neighbours by collagen allows great precision in the control of contraction by the nervous system. The fast, controlled response of skeletal muscle makes possible the eye movements that follow flight of an insect, the finger movements of the virtuoso pianist and the delicate combinations of movements of tongue, lips and palate needed to produce speech.

But how can we understand the precisely graded and deliberately slow contractions of muscles in, say, the hand as we stretch it out to stroke a cat? Slow and gentle movements seem incompatible with the rapid and powerful response of a muscle fibre. There are two answers. If we think first of the single fibre, its responses to a widely spaced series of nerve impulses will be a number of minute contractions separated by periods of relaxation; its response to a volley of very closely spaced impulses will be continued and increasing contraction up to a maximal shortening. So grades of response are possible in a single fibre. At a higher level of organization, a muscle is supplied by many nerve fibres, each of which innervates a small number of muscle fibres. An impulse in a single nerve fibre will cause contraction in only a few of the muscle fibres making up the muscle. Stimulation of more and more fibres gives increasingly powerful contraction of the muscle as a whole. In similar fashion, a muscle can hold a contraction for a long time, although the individual fibres fatigue fast, by rotating the contraction between groups of fibres.

The recognition of striated muscle

The long, regular cylinders of striated muscle, with their obvious cross-striations, cannot easily be mistaken for anything else (Fig. 9.6): striated muscle fibres are longer and wider than smooth muscle cells, with very few exceptions. Since striated muscle fibres are regular cylinders, they do not vary much in cross-sectional diameter (Fig. 9.10), unlike smooth muscle cells. Whether seen in longitudinal or transverse section, the nuclei appear small, and are peripherally placed under the sarcolemma, unlike those of smooth muscle.

Cardiac muscle

The third variety of muscle is cardiac muscle, found only in the wall of the heart, and, in larger mammals such as ourselves, in the walls of the great veins as they approach the heart. Cardiac muscle has many of the characteristics of striated muscle, with the same organization of contractile proteins into sarcomeres. Unlike striated muscle, it is composed of individual cells rather than large, multinucleate cylinders, and possesses elaborate junctions between neighbouring cells.

Seen in the light microscope (Fig. 9.11), the cells of cardiac muscle are short and fat, and often clearly marked off from the cells at either end by a thick line, the intercalated disc.

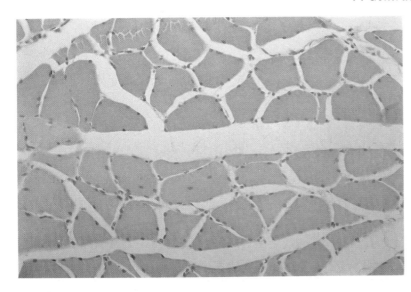

Fig. 9.10 Transverse section of striated muscle. H & E. (× 200.)

There may be one or two nuclei per cell, which are quite large and centrally placed in the fibre: the cross-striations are clearly visible. These cells often branch, forming stubby Y-shapes when separated from each other. The striated appearance implies a rapid contraction and relaxation, and inability to hold a contraction for long without fatiguing. With the TEM, the nucleus is seen to have a small cap of cytoplasm at each end containing RER and SER, a Golgi apparatus and many mitochondria. Mitochondria lie between the myofilaments in great numbers (Fig. 9.12), and the sarcoplasmic reticulum and T-tubule system are again present, forming triads.

The intercalated disc

Cardiac muscle differs from skeletal in its junctions with neighbouring cells — or rather, with the cells at either end. The thick, straight line of the intercalated disc, as seen with the

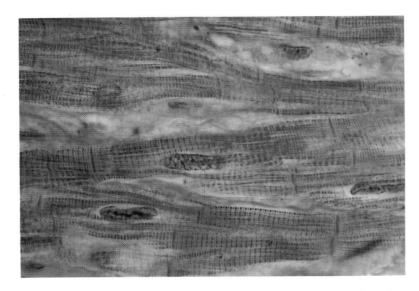

Fig. 9.11 Longitudinal section of rat cardiac muscle. PTAH. (× 960.)

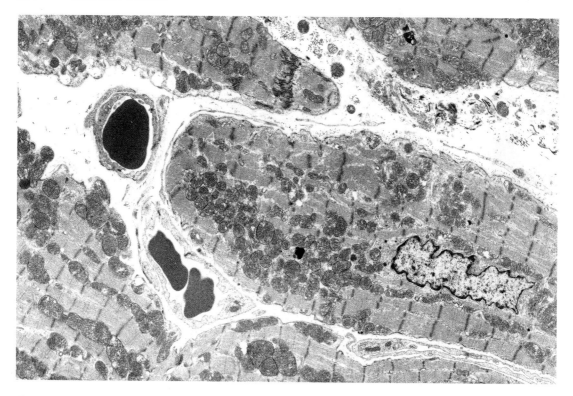

Fig. 9.12 TEM of human cardiac muscle fibres with connective tissue and capillaries between them. (× 5000.) (Courtesy, Dr D. Haynes.)

light microscope, becomes a very elaborate and irregular junction with the higher resolution of the TEM (Fig. 9.13). Many desmosomes can be recognized in the disc, together with areas of membrane resembling the intermediate junction of epithelial cells, called here the fascia adherens.

What are the characteristics of the intermediate junction, or zonula adherens? (Note 9.K)

The irregular junction and high density of desmosomes and intermediate junctions can

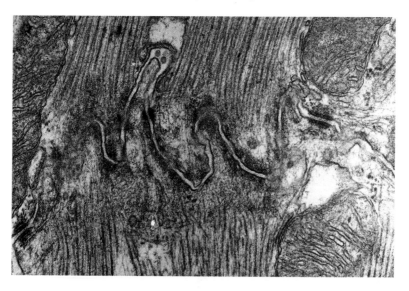

Fig. 9.13 TEM of an intercalated disc between two cardiac muscle fibres of the rat. (× 47 000.) (Courtesy, Dr B. Gannon.)

only imply that the membranes concerned are specialized to resist pulling apart.

One other specialization occurs at the intercalated disc, though not obvious on sectioned material, and that is a large number of gap junctions.

The functional characteristics of cardiac muscle

The branching, end-to-end arrangement of cardiac muscle cells results in a continuous basket-like mass of muscle around the cavities of the heart. When these cells get shorter and fatter, the effect is to reduce the volume of the central cavities, squeezing blood out of the heart.

Cardiac muscle cells have the ability to contract regularly, without external stimulus, rather like the spontaneous contractions of smooth muscle. It is possible to harvest isolated cells from the heart of an experimental animal and observe them in culture, and it is fascinating to watch them contracting rhythmically, in the absence of any nerve supply or contact with other cells. When such cells are linked by gap junctions, contraction in one will stimulate contraction in those at either end. Effectively, the cell with the fastest rate of spontaneous contraction will drive the others, imposing its rhythm on the whole cell mass.

The recognition of cardiac muscle

Cardiac muscle should not be difficult to recognize. The cross-striations and intercalated discs give the cells a very different appearance from smooth muscle. The irregular, branching cells, central nuclei and intercalated discs contrast with the large, regular cylinders and peripheral nuclei of striated muscle.

The biological importance of muscle contraction

These three proteins, actin, myosin and alpha-actinin, are present through a very wide range of cells, muscle and non-muscle alike. The shortening produced by their interaction in the presence of ATP and Ca^{2+} pulls nearer together the sites where actin is anchored to alpha-actinin, and this reaction is the basis of almost all cell movement. The only exception is movement based on microtubules, which is responsible for transport within the cell, the beating of cilia and the swimming of sperm.

Just about all our interactions with the surrounding world are due to the contraction of muscle. Talking, chewing, walking, writing, fighting, using tools are all possible only through the interactions of these three molecules. Everything that human beings have created, from flint arrowheads to ballistic missiles, from symphonies to gardens, from loaves of bread to machine tools have come into being through the shortening of muscle cells.

Further reading

Goldspink, G. (1980). Locomotion and the sliding filament mechanism. *In* "Aspects of Animal Movement", Elder, H.Y. and Trueman, E.R. (Eds) pp. 1–25. Cambridge University Press. Clear recent summary.

Huxley, H.E. (1965). The mechanism of muscular contraction. *Scient. Am.* **213:6**, 18–27. Review of the work that sorted out the structure of striated muscle.

Lazarides, E. and Revel, J.-P. (1979). The molecular basis of cell movement. *Scient. Am.* **240:3**, 98–107. Excellent review on contraction in non-muscle cells.

Murray, J.M. and Weber, A (1974). The co-operative action of muscle proteins. *Scient. Am.* **230:2**, 58–71. Reviews the mechanism of muscle contraction.

Schoenberg, C.F. and Needham, D.M. (1976). A study of the mechanism of contraction in vertebrate smooth muscle. *Biol. Rev.* **51**, 53–104. Summarizes the data on smooth muscle structure and function.

10
Harnessing contraction to produce movement

The ability of muscle cells to shorten is only the first stage in producing effective movement. That shortening must be harnessed and converted into the movement of one structure relative to others. The mechanisms which have evolved in the body to transduce the contraction of muscle into useful movement, with one exception — the calcification of bone — are developments of features in connective tissue which you already know from Chapter 5. The extracellular components of connective tissue form the basis of specialized structures with a surprising range of functional characteristics.

What are the extracellular components of connective tissue? (Note 10.A)
Collagen fibres strongly resist stretching, but can readily be coiled or crumpled. They transfer the shortening of muscle to the rigid plates and levers of the skeleton. In addition, they form the basis of the skeletal tissues, cartilage and bone. These tissues, possessing tensile strength, which is largely due to their collagen centent, also resist compression and shearing forces, by adding to the collagen fibres a matrix which prevents crumpling. In cartilage, this matrix is composed of GAGs in large complexes with protein; in bone, the collagen is made rigid by depositing inorganic salts of calcium and phosphorus on its surface.

Movement of any structure in the body must affect adjacent structures. This chapter will end with a look at the various ways in which the displacement of one structure relative to another is accommodated.

Tendons

We have already seen the fine network of collagen that surrounds each fibre of striated muscle, and noted the thicker layers of collagen around bundles of fibres and whole muscles. At the ends of a muscle these collagenous streaks merge into tendons, made up of collagen fibres densely packed, and parallel to the long axis of the muscle (Fig. 10.1). Tendons vary greatly in their size and arrangement. Some are very short indeed, so that a muscle seems to attach directly to bone; some are long cords, attached to bone a considerable distance from the muscle, sometimes travelling around pulleys to redirect the axis of pull of the muscle; some again are thin, flat sheets of collagen. In every case, the high tensile strength of collagen is used to transfer the shortening of muscle into the movement of some structure, usually some part of the skeleton.

The dense bands of parallel collagen fibres which form tendons (Fig. 10.1) contain

(a)   (b)

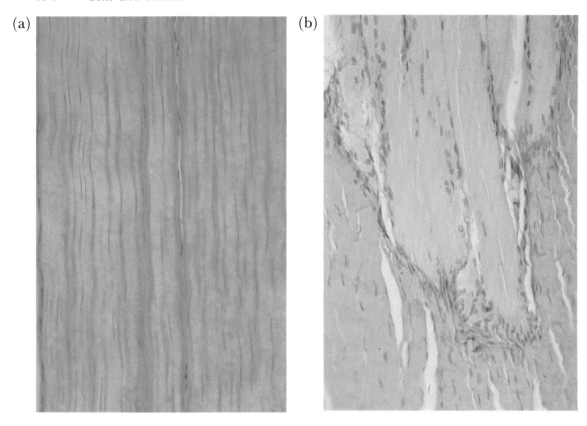

Fig. 10.1   (a)   Longitudinal section of a tendon. H & E. (× 200.) (b)   Junction of striated muscle (top) with tendon (below). H & VG. (× 200.)

relatively few fibroblasts, and these do not look very active histologically. They appear on sections as thin, flattened nuclei sandwiched between the collagen fibres. A few, small blood-vessels can be seen, running parallel to the collagen, and GAGs are present between the fibres. Such tendons have surprising tensile strength — usually estimated at about 9 kg/mm². At this figure, the Achilles tendon of man (just behind the ankle) should be able to support the weight of a small car such as the Mini.

Smooth and cardiac muscle is usually arranged in a woven fashion around a central cavity. Here, too, a fine network of collagen surrounds the muscle fibres, and thicker layers of collagen often invest the sheet of muscle, converting the shortening of individual fibres into a compression of the central cavity.

Cartilage

But tendons can be crumpled. Resistance to compression can only come about through embedding the collagen fibres in a firm matrix of some sort to prevent crumpling. In cartilage, this is achieved by surrounding the collagen fibres with complexes of GAGs with protein.

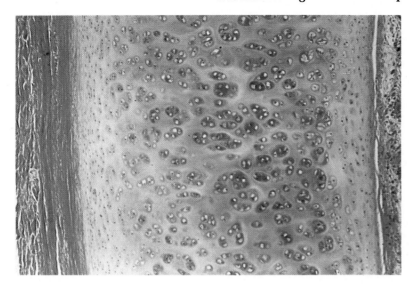

Fig. 10.2   Hyaline cartilage from the wall of the trachea. Masson's trichrome. (× 38.)

What are GAGs? What are their chief characteristics as a component of connective tissues? (Note 10.B)
Cartilage is semi-transparent, with a blue-grey, shiny cut surface. It is tough and springy — "gristle" — and often difficult to separate cleanly from the surrounding, whitish connective tissue. One cannot easily identify the bundles of collagen histologically, particularly in the form of cartilage known as hyaline (Fig. 10.2). Their presence and orientation can be demonstrated, however, by a number of techniques. The simplest is to stick a fine needle, dipped in indian ink, into the surface of the cartilage. Such a puncture usually produces a slit-like hole, with its long axis in the major direction in which the collagen is oriented.

Cartilage often occurs in relatively small nodules or thin sheets in the body, in which case it is avascular. The diffusion of oxygen and nutrients through cartilage appears to be rapid and relatively easy. Larger volumes of cartilage, such as occur at the ends of long bones in a young child, have blood-vessels which penetrate into the cartilage from the surface in canals of connective tissue; these canals also permit the entry of cells during ossification — the replacement of cartilage by bone during growth.

The matrix of cartilage

Cartilage develops within connective tissue, when a group of cells otherwise indistinguishable from fibroblasts starts to produce the specialized matrix which fills the extracellular spaces around them. Not surprisingly, then, cartilage is surrounded by connective tissue, and there is no abrupt edge to the cartilage, but, rather, a gradual transition from connective tissue (Fig. 10.3). As the cartilage increases in size with the laying down of more and more extracellular material — the cartilage matrix — the cells become quite widely separated from each other. Often, the surrounding connective tissue becomes compressed by the increasing volume of the nodule of cartilage, forming a collagenous capsule around it. The two major components of the matrix are the extracellular fibres, collagenous and elastic, and the GAGs in which they are embedded.

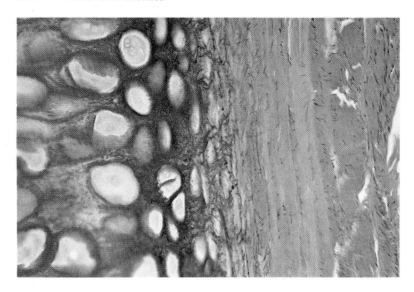

Fig. 10.3 The edge of an elastic cartilage (left), merging into collagenous connective tissue (right). Orcein stain. (× 405.)

Proteoglycan aggregates. In cartilage, the GAGs, synthesized by the cartilage cells, form extremely large aggregates in the extracellular spaces. Each aggregate has a linear backbone of hyaluronic acid, to which are attached a large number of protein side-arms; these in turn carry large numbers of molecules of chondroitin sulphates and of keratin sulphate. Remembering that both of the latter carry a series of negative charges, it is not surprising that each part of this huge molecular aggregation attempts to repel every other part. The result is that the whole aggregate looks like a bottle-brush and is springy, resisting any attempt to deform or crush it.

The GAGs of connective tissue are molecules with large domains, holding a great deal of the extracellular water effectively bound; the much larger proteoglycan aggregates of cartilage do the same to an even greater degree.

Proteoglycan aggregates and collagen. The collagen fibres running through cartilage are kept under tension all the time by the large proteoglycan aggregates which surround them. Each aggregate is attempting to open out fully and to occupy the full volume of extracellular fluid which it can theoretically fill; the collagen fibres surrounding it resist this expansion. The result is an equilibrium. A deforming force, such as compression due to weight-bearing, can supply enough energy to overcome the electrostatic repulsion within the proteoglycan aggregates locally, crumpling them further and altering the shape of the piece of cartilage. As this deformation proceeds, the increasingly crowded negative charges on the GAGs resist further compression more and more strongly. When the force is removed, the aggregates spring back to their original shape. The effect is to produce a tissue that is tough and springy, with a considerable resistance to compression.

As in ordinary connective tissue, the collagen fibres are not arranged at random, but have definite orientation to give the cartilage the greatest resistance to the forces normally applied to it. Each cartilage cell is surrounded by a spherical basket of collagen, in addition to the larger and straighter bundles of fibres running through the matrix. Note that much of the volume of cartilage is occupied by water, held bound by the GAGs of the proteoglycan aggregates. Under a load which only lasts a very short time, little change in volume occurs in cartilage and recovery to its original shape takes place almost instantaneously on removing the load. But if the load continues for minutes or hours, continued compression of proteoglycan aggregates forces out of them some of the bound water. In these conditions, water can be squeezed out of the cartilage, and, on removing the load, recovery has two phases — an initial springy rearrangement of the compressed aggregates followed by a slower uptake of water and swelling back to the original shape and volume.

As in ordinary connective tissue, elastic fibres may be present in cartilage, reinforcing its ability to return to the original shape on removing a deforming force.
Given the electrostatic charge carried by proteoglycan aggregates, how would you expect the matrix of cartilage to stain with haematoxylin and eosin? (Note 10.C)

The cells of cartilage

The cells of cartilage have two names, chondroblast and chondrocyte. Chondroblasts are actively secreting new matrix, whereas chondrocytes are surrounded by mature matrix and are relatively inactive. All stages in the transition from apparently normal fibroblasts to active chondroblasts can be seen at the edge of a nodule of growing cartilage (Fig. 10.3). As more and more extracellular matrix is laid down, chondroblasts become more widely separated from each other, lying in small spaces or lacunae in the matrix (Fig. 10.2). Chondroblasts retain the ability to divide after they begin to synthesize matrix, so that lacunae are frequently clustered in pairs or groups of four in mature cartilage.
Given the synthetic abilities of chondroblasts, predict their appearance by TEM. (Note 10.D)
Note the large size and rounded shape of the chondroblast. The chondrocyte is rather similar in appearance, but contains less of the synthetic apparatus so evident in the chondroblast. Chondrocytes in mature cartilage may lie a considerable distance from the nearest capillary. Much of the volume of the matrix of cartilage is water: diffusion of materials through this matrix is surprisingly rapid, provided the molecule is small and does not carry a net positive charge.

The recognition of varieties of cartilage

I have written so far as if all cartilage is similar. In fact, just as there are local variations in the relative proportions of cells and extracellular materials in connective tissue, so are there considerable differences in the composition of cartilage from place to place in the body. This sort of variation is continuous, but the extremes of variation have been given different names by histologists.

Hyaline cartilage. This is the commonest form of cartilage, particularly in the young animal, in which it forms the precursors of many of the bones. In the adult, it persists in the walls of the respiratory tract, forming the larynx, or voice-box, as well as a large number of C-shaped bars and flat plates which hold open the trachea and bronchi. It joins the anterior ends of the ribs to the sternum, forming the costal cartilages, and, as we shall see later, covers the surfaces of bones at synovial joints.

The name, hyaline, means glassy. Both in blocks viewed naked-eye and in histological

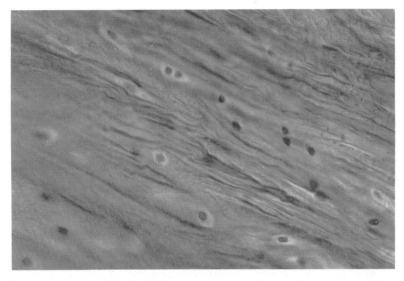

Fig. 10.4 Fibrocartilage from the insertion of a large tendon into bone. Dense collagen can be seen, as well as chondrocytes in lacunae. H & VG. (× 505.)

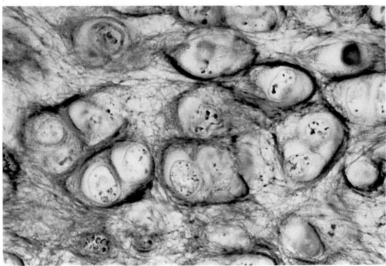

Fig. 10.5 Elastic cartilage from the pinna of the ear. Verhoeff's stain. (× 385.)

sections, it looks uniform and translucent, and it comes as a surprise to students to learn that about 40% of its dry weight is collagen. The material in which it is embedded has a very similar refractive index to that of collagen. Its histological appearance is quite characteristic, with the uniform basophilic matrix and the scattered lacunae containing chondrocytes. The gradual transition from fibrous connective tissue to cartilage is another characteristic feature (Figs 10.2, 10.3).

Fibrocartilage. Where the collagenous content of the matrix is much higher than in hyaline cartilage, the tissue is called fibrocartilage. This is found in the tough intervertebral discs and other intra-articular discs, and frequently occurs where large tendons or ligaments approach their insertion into bone. The high collagenous content gives it a whiter colour naked-eye, and the wavy bundles of collagen can be clearly seen histologically (Fig. 10.4). What distinguishes it from the dense fibrous tissue of tendons and ligaments (Fig. 10.1) is the presence of small clusters of rounded chondrocytes in lacunae between bundles of fibres: these look very different from the flattened nuclei of fibroblasts in a tendon, and they should alert you to the presence of proteoglycan aggregates in between the collagen fibres and the cells.

What effect would this high proportion of collagen to proteoglycan aggregates have on the response of fibrocartilage to deforming forces? (Note 10.E)

Elastic cartilage. The very springy skeleton of the outer ear, the nasal cartilages and the epiglottis contains a high proportion of elastic fibres, embedded with the collagen in proteoglycan aggregates. With haematoxylin and eosin this can be confused with hyaline cartilage. Special stains for elastic fibres, however, reveal the dense network throughout the cartilage (Fig. 10.5).

Ageing and repair in cartilage

Cartilage is normally subjected to repeated deforming forces, and it is relatively avascular, as we have seen. Perhaps it is the combination of these two factors which makes it a tissue that causes a great deal of trouble as the individual ages. Diffusion of oxygen and nutrients through the matrix is sufficient to keep chondrocytes alive in normal, healthy cartilage, but is probably not enough to support the processes of cell division and active synthesis needed to repair damaged cartilage. The ingrowth of connective tissue, carrying blood-vessels and the fibroblast-like precursors of chondroblasts, is often incompatible with the forces normally applied to cartilage, which are sufficient to close capillaries.

One further aspect of the ageing of cartilage is its tendency to accumulate calcium, presumably by ionic binding to the GAGs. Calcified cartilage loses most of its springiness and is, instead, rigid and brittle. The costal cartilages are twisted by the muscular action of breathing in, and normally spring back when this action ceases, contributing to the action of breathing out. Calcification in these cartilages produces a rigid rib cage, incapable of expansion and contraction.

Bone

Bone, like cartilage, is built around collagen, but there the likeness ends. Whereas cartilage achieves resistance to deforming forces by packing large proteoglycan aggregates between the bundles of collagen, producing a tough but flexible tissue, bone involves the deposition of inorganic salts of calcium and phosphorus along and between the collagen fibres. Bone is rigid and hard. It resists compression as well as tension while remaining surprisingly light. The average breaking stress of the rat's femur is similar to that of cast iron, yet a comparable volume of cast iron is three times heavier than bone.

But bone is not just a rigid material capable of withstanding stress: it is a living tissue, capable of remodelling itself to meet changing stresses and of repairing itself following fracture.

The basic unit of bone is the spicule which is essentially a bundle of collagen fibres, given resistance to crumpling by the deposition of calcium phosphate around and between the

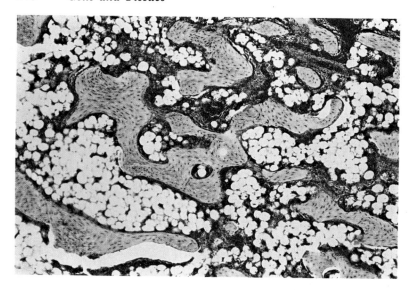

Fig. 10.6 *Spicules of bone (yellow) surrounded by fat cells in the bone marrow. H & VG. (× 40.)*

fibres (Fig. 10.6). Such spicules may run as thin trabeculae through connective tissue, with capillaries always close and a lining of cells on their surfaces capable of laying down more bone or of resorbing existing bone. In positions that require great resistance to stress, like the shafts of long bones, such spicules may be laid down tightly packed: here the need to have capillaries near is met by arranging the spicules concentrically around a central blood-vessel (Fig. 10.7). Numbers of such concentric patterns together form compact bone. The very vascular organization of bone contrasts strongly with the structure of cartilage, and contributes to its ability to remodel and respond to changed stresses, and to repair itself.

The extracellular matrix of bone

Bundles of collagen fibres form the basis of bony spicules, whether in isolated trabeculae or in compact bone. The spaces between the fibres contain a small amount of GAGs, but bone, a rigid tissue, has no elastic fibres. Along and around each collagen fibre, calcium phosphate is deposited in crystalline form resembling the mineral hydroxyapatite.

The deposition of calcium phosphate has stimulated a great deal of research. For a salt to crystallize from solution, certain criteria must be met. The concentrations of calcium and phosphate must exceed a given level: in other words, the solution must be saturated. At such a concentration, a crystal of calcium phosphate will grow by the addition to it of ions from solution. But to initiate the first crystal requires even higher concentrations of calcium and phosphate. A tiny crystal has a large surface area relative to its volume, and has a high probability of losing ions into solution. So a supersaturated solution is needed to start the process of crystal formation, while crystal growth can proceed at lower concentrations once the seeding crystals reach a given size. Tissue fluids such as blood plasma can support crystal growth, but are not able to initiate crystallization.

The hypotheses that have been proposed, tested and rejected over the last 15 years to account for the deposition of calcium phosphate in bone are well described by Simkiss (1975) and by Ali (1980). The present hypothesis is that cells involved in calcification bud

Fig. 10.7   Transverse section through an Haversian system in compact bone. A ground section of bone, viewed by interference contrast. The lacunae containing osteocytes are black: the concentric rings of collagen at different orientations produce alternating dark and light rings. (× 400.)

off tiny membrane-bound structures called matrix vesicles. These contain high concentrations of alkaline phosphatase, and probably a mechanism for either pumping calcium ions into their interior, or ion channels in the membrane permitting the entry of calcium by diffusion. Phosphate groups are split from organic compounds such as ATP by the phosphatase, producing sufficiently high concentrations of calcium and phosphate within the vesicles to initiate crystal formation. As the crystals grow, the vesicle is ruptured, and further crystal growth occurs from the levels of calcium and phosphate in the local extracellular fluid. Matrix vesicles provide a small membrane-bound environment in which the conditions for intitial crystal formation can occur. It is possible that the collagen of bone differs from collagen elsewhere in favouring the growth of crystalline calcium phosphate over its surface, once crystallization starts.

The cells of bone

Much of the process of bone deposition occurs at the surface of bone spicules, and it is here, too, that the removal of bone takes place. These two processes are associated with two cell types: osteoblasts and osteoclasts. As bone grows, an occasional osteoblast becomes surrounded by it; in its new position, continued deposition of bone becomes impossible — the cell ceases to be actively synthetic. It is now called an osteocyte. Let us examine these three cell types in more detail.

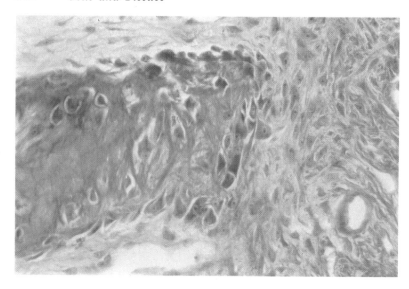

Fig. 10.8   A spicule of bone (stained blue), surrounded by active osteoblasts (deep purple).   Mallory's trichrome. (× 405.)

Osteoblasts. These (Fig. 10.8) are rather plump cells, often chunky, arranged in an irregular row along one or more surfaces of a bony spicule. They are very actively synthetic, making collagen and GAGs as well as the enzymes need for bone deposition, which are budded off the cytoplasm in matrix vesicles. Their intense cytoplasmic basophilia indicates a high content of RER and free ribosomes. By now you should have no difficulty in predicting their appearance by TEM.

Osteocytes. These lie in lacunae in the bone matrix and are smaller than the osteoblasts from which they were derived. Unlike chondrocytes, they are not completely isolated in the matrix, but have many fine, cytoplasmic processes reaching out in tiny channels in the bone and making contact with the processes of neighbouring osteocytes. Remember that the matrix of cartilage is freely permeable to oxygen in solution and to many small molecules, so that the provision of supplies to chondrocytes is assured. In bone, the matrix is very different, and it seems that these tiny channels (Fig. 10.7) are a mechanism for providing access to necessary materials as well as possible cell-to-cell communication.

Osteocytes may well be capable of acting on the large surface of bone matrix to which they are exposed. Withdrawal of calcium and phosphorus from the skeleton can be very rapid in certain experimental conditions, and it does not appear to occur solely at the exposed surfaces of bony spicules.

Osteoclasts. These (Fig. 10.9) are very large, multinucleate cells, closely applied to the surface of a bony spicule and often lying in a shallow depression in its surface, known as Howship's lacuna.

There are surprisingly few osteoclasts on any section of bone, but the chances of finding them are much better if one looks at a surface of bone which is being rapidly resorbed. The inner surface of skull bones in a growing animal or the bone surrounding a growing tooth germ are good exmples. But one comes to the conclusion that bone deposition always exceeds bone resorption if the numbers of osteoblasts and osteoclasts on any section are compared. It is very likely that other, mononuclear cells — perhaps osteoblast-like, or, as we have noted, osteocytes, perhaps tissue macrophages — may be involved in bone

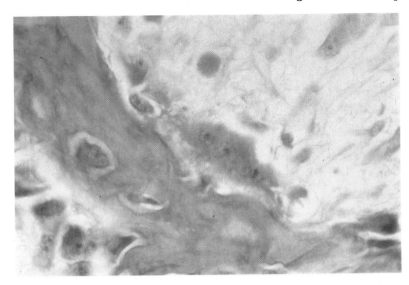

Fig. 10.9   A spicule of bone (stained blue), with a large, multinucleate osteoclast at its surface (centre). Mallory's trichrome. (× 1015.)

resorption also, and that osteoclasts are specialized for the very rapid, local removal of bone matrix.

The surface of the osteoclast that faces the bone forms a very complex series of infoldings of cell membrane, called a ruffled border. Many small membrane-bound vesicles are associated with the cytoplasmic side of these infoldings. There is probably a two-way traffic at this surface, with the secretion of enzymes into the extracellular space and the phagocytosis of fragments of matrix. The cytoplasm contains many mitochondria, particularly near the ruffled border, and many primary lysosomes, together with the RER and Golgi apparatus associated with lysosome synthesis.

The micro-architecture of bone

We have considered bone so far as thin spicules of calcified matrix. How is this material arranged to build up the structures we recognize as bones?

There are two distinct patterns of bone, cancellous and compact, and both are present in any bone.

Cancellous bone. This consists of branching and anastomosing plates and spicules of bone, often organized into elaborate patterns. These are difficult to appreciate on histological sections, where they appear as irregular shapes, but they can be well seen in radiographs of thick sections of bone (Fig. 10.10). These patterns have been analysed in terms of the stresses applied to particular bones in life, and the bony trabeculae fall into two groups. Some are oriented to resist compressive forces applied to the bone, others clearly resist tensile forces, and the two types of trabeculae usually cross at about 90°. The economy with which bony trabeculae are arranged along stress lines in cancellous bone is extraordinary, producing a light yet very strong structure.

This arrangement of bone permits capillaries to come very close to the cells at the surfaces of the bony trabeculae. Histologically, some of these surfaces will be lined by a row of osteoblasts, indicating the deposition of new bone. The cells are separated from the fully calcified matrix by a thin layer of collagen which is as yet imperfectly calcified. Many of the

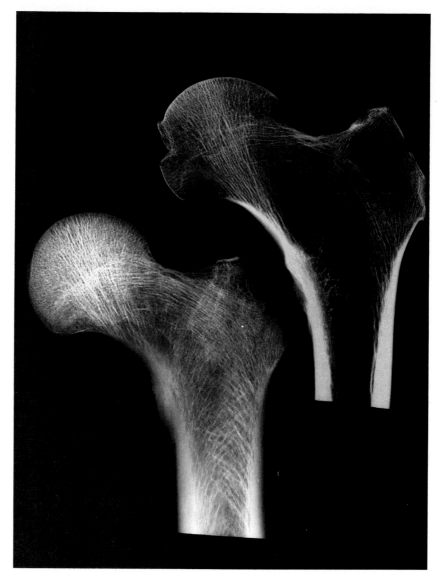

Fig. 10.10   A radiograph of two sections through the upper end of a human femur. Dense cortical bone, and criss-crossing trabeculae of cancellous bone are clearly seen.

surfaces of the irregular spicules of the bone seen on section will have rather flat, inactive cells which resemble fibroblasts applied to them. Occasionally, a large, multinucleate osteoclast will be seen, usually on the opposite surface to a row of active osteoblasts.

Cancellous bone occurs within bones of all shapes. In the long bones of the limbs, there is more cancellous bone at the ends and relatively little in the shaft. Between the trabeculae lies connective tissue with blood-vessels. In some bones, particularly in the sternum, vertebral bodies and innominate bone in the adult, red or active bone marrow lies in these spaces, the site of production of red and white blood cells. In other bones, the marrow is normally inactive in the adult and is largely fatty connective tissue, or yellow marrow. In the infant and child, red marrow is much more extensive.

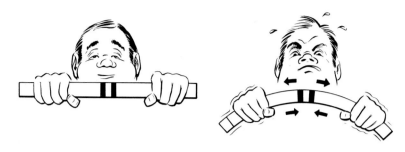

Fig. 10.11 Deforming a cylinder compresses its concave surface while attempting to stretch its convex surface.

Compact bone. If a cylindrical bar is bent by a deforming force (Fig. 10.11), one surface of the cylinder, on its concave side, will be compressed, whereas the opposite surface will be under tension. In the centre of the cylinder, these forces, which are acting in opposite directions, will tend to be minimal. The most economical design to resist bending is to have the greatest strength at and just below the surface of the cylinder and to have little or no resistance in its centre. This is the structure of every bone. Compact bone, essentially consisting of many trabeculae fused into a solid mass, forms the outer plate, leaving the interior to be buttressed by trabeculae of cancellous bone. In the long bones of the limbs resistance to bending is particularly important. Here the outer rim of compact bone can be very thick in the femoral and tibial shafts, and the centre of the shaft is almost free from bone. In flat bones such as the scapula or the bones of the vault of the skull, the plates of compact bone are thinner and the proportion of cancellous bone higher; the same is true of the ends of long bones.

Compact bone is organized in units, called osteons (Fig. 10.7), surrounding a central blood-vessel. The cylinder of bone matrix around a blood-vessel can also be called a Haversian system. Around the central blood-vessel lie concentric cylinders of collagen forming spirals of varying pitch, all of the layers being calcified. Between the layers lie osteocytes, each in its lacuna and linked to its neighbours, and ultimately to the extracellular space around the blood-vessel, by minute canaliculi (Fig. 10.7). Such Haversian systems have a rather constant maximum diameter, presumably limited by the ability of the blood-vessel to nourish cells at a distance through canaliculi.

Haversian systems are cylindrical, yet they form a solid mass of bone. How are the cylinders packed together? In the spaces between complete, cylindrical systems lie the incomplete remnants of other, older systems.

The plasticity of bone. Bone is the body's store of calcium and phosphate, in some equilibrium, not necessarily direct and simple, with the levels of these ions in the blood and tissue fluids. Measurements with radioactive isotopes of phosphorus and calcium have demonstrated just how rapid and continual the interchange of ions is. It has been calculated that, out of the total calcium ions in the blood, one in four exchanges with bone calcium every minute.

It is possible to conceive of such rapid exchange occurring without any change in the organization and pattern of bone, just by ions leaving and joining the crystals of hydroxapatite at their surfaces. The histological picture of bone that has already been described suggests, however, that remodelling of the trabeculae of bone is constantly taking place. In cancellous bone, surfaces lined with osteoblasts are sites of deposition of new matrix, whereas the Howship's lacunae with osteoclasts represent the removal of old matrix. In compact bone the presence of remnants of old Haversian systems in the spaces between complete systems also suggests a process of removal and renewal.

Autoradiography after the injection of radioactive calcium or phosphorus demonstrates a fine, apparently random incorporation of radioactivity into the surfaces of trabeculae in cancellous bone, and into the centre of Haversian systems of compact bone. In addition, much more massive incorporation occurs on particular surfaces of trabeculae, and into complete, new Haversian systems within compact bone.

Bone is being constantly remodelled, with the removal of some bits of matrix and the deposition of new. Remodelling is particularly evident if the stresses placed on the bone change, following the out-of-line repair of a fracture, for instance. I have seen radiographs of the tibia which showed apparent remodelling after a girl started to wear high heels for the first time. Not only does the pattern of bony trabeculae change in response to new stresses, but the overall density of bone, its degree of calcification, varies with the load placed on it. One of the problems faced by astronauts is loss of bone matrix as a result of prolonged weightlessness.

The periosteum

At the surface of every bone lies connective tissue, continuing out into the rest of the interior of the body. What limits ossification, preventing the spread of calcification?

The surfaces of bone (with the exception of joint surfaces, which will be considered later) are covered with a tissue called the periosteum. This has two components: an inner layer, rich in cells which are capable of producing bone, and an outer, tough layer of collagen. At times, rupture of the periosteum may lead to the escape of bone-forming cells into surrounding tissues, resulting in the ossification of a tendon or muscle.

This tough, fibrous sheath, the periosteum, is firmly attached to the bone beneath by a large number of bundles of collagen which leave it at right angles to enter the bone, becoming calcified as they form part of the compact cortical bone.

Attachment of tendons to bone. Where a tendon attaches to bone, bundles of collagen splay out to fuse with the tough, fibrous periosteum. Other bundles penetrate the periosteum to continue into the compact cortical bone, becoming calcified as they do so. Some bundles fan out beyond the compact bone to continue on as trabeculae of cancellous bone.

It is misleading to talk of the attachment of a tendon to bone. The one really is continuous with the other (Fig. 10.12). Collagen is the basis of both tissues, and the bundles of collagen continue on from the tendon into the bone. It is very unusual to tear a tendon or ligament from its bony attachment: it is more common to rupture the tendon or to fracture the bone.

We have already mentioned the presence of fibrocartilage in large tendons as they approach their insertions.

Effects of movements on surrounding structures

Movement of any organ in the body involves some displacement of surrounding structures,

Fig. 10.12   The attachment of a tendon to bone. Collagen fibres, passing through a layer of fibrocartilage (centre), continue on into bone (above). H & VG. (× 40.)

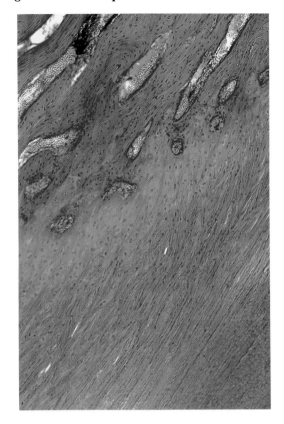

and this is particularly true of the relatively large movements of bones in locomotion. Loose connective tissue is finely varied in its organization to accommodate some movements. As we have seen in the skin, planes often exist, such as that between the dermis and underlying structures, which consist largely of extracellular fluid and GAGs, permitting some degree of independent movement of muscles beneath relative to skin. *What limits the extent of such movements? (Note 10.F)*

There is a limit to the movement possible in loose connective tissue. Two mechanisms have evolved to accommodate greater ranges of movement than are possible in the network of fibres, cells and capillaries of connective tissue — these are pads of fat and fluid-filled spaces.

Fat

This is one of the normal constituents of connective tissue, but it has several characteristics that make it peculiarly useful in accommodating movement. The first is its intracellular storage as a single, large droplet. Since fat is fluid at body temperature, a mass of fat cells (Fig. 5.6) can be deformed readily. Secondly, fat as a tissue has a relatively low metabolic rate and a correspondingly low blood supply: as we have seen above, it is blood-vessels and their surrounding collagen that often limit movements in connective tissue. So a large cluster of fat cells, with relatively few vessels entering at one surface only, is capable of adapting to a far wider range of movement than other varieties of connective tissue.

Pads of fat are frequently found near joints and around abdominal organs that move. In the bony orbit, the space behind the eyeball is filled with fat: this adjusts to the contractions of the muscles that move the eye, and also cushions inward movements of the eyeball due, for instance, to blows on the eye.

Fluid-filled sacs

But even pads of fat can only accommodate so much displacement. Sliding or shearing movements of one structure on another are particularly difficult to adjust to without putting unacceptable stresses on blood-vessels and their supporting collagen. The body has evolved a recurring pattern to meet such demands, the formation of a fluid-containing sac between the moving surfaces. From such a sac, cells, extracellular fibres and blood-vessels are excluded by the lining membrane, leaving only a thin film of fluid rich in GAGs.

The muscles that flex the fingers are large and powerful. They cannot be positioned in the hand without restricting movement, so they form a bulky mass in the forearm and their contraction is transferred to the bones of the fingers by long tendons which pass in front of the wrist and across the palm of the hand. If you watch your fingers as they flex, it is clear these tendons move several centimetres past the bones around the wrist, which remain stationary. These tendons are surrounded at the wrist by fluid-filled sacs, called tendon sheaths or synovial sheaths.

The chest wall moves upwards and outwards on breathing in, whereas the surface of the lung in contact with it moves downwards and outwards, following movement of the diaphragm as well as that of the chest wall. The shearing movements between lung and chest wall are much greater than can be accommodated by connective tissue, so the lung is surrounded by a fluid-filled sac, the pleural cavity, from which cells, fibres and blood-vessels are excluded. Similar cavities enclose the heart and the abdominal contents — the pericardial and peritoneal cavities respectively.

In histological material one usually sees only one wall of such a sac on the outer side of some structure or organ (Fig. 11.8a). The cells lining the sac are flattened, squamous, only one layer thick. Beneath this layer lies connective tissue and blood-vessels. Normally, only a thin film of fluid separates the two surfaces of such a sac, allowing the smooth surfaces of flattened cells to move freely across each other.

Joints

Joints are the places where skeletal elements — bones or cartilages — meet. In some positions in the body, joints hold bones together with the minimum of movement; in others, joints are specialized to permit a wide range of movements. Over this spectrum of mobility, there are, inevitably, variations in the structural organization of joints. We have by now met all the tissues which take part in the different types of joint, so let us use our existing knowledge to predict the histological structures of joints.

Joints with little or no movement

The vault of the skull is made up of several bones, which do not move significantly relative to one another. The lines along which one such bone makes contact with its neighbours is, by definition, a joint. These lines are not straight, but very irregular, like some elaborate jigsaw puzzle.

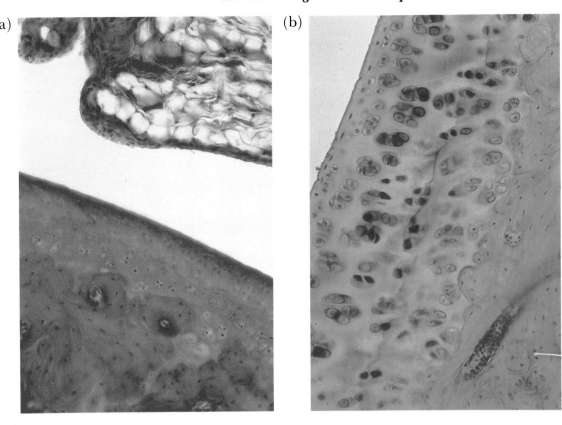

Fig. 10.13   Tissues at a synovial joint. (a) Synovial membrane (above) and articular cartilage (below). H & VG. (× 100.) (b) Articular cartilage lines the joint cavity (top left), with bone deep to it (bottom right). Carmine. (× 200.)

What is the simplest arrangement of tissue elements which would join these bones in such a way as to prevent movement? (Note 10.G)

The joint in the midline anteriorly between the two halves of the pelvic girdle is again one at which very little movement occurs. There is no active movement here, in the sense that no muscle is attached so that its contraction can move one half of the pelvis relative to the other. But minor degrees of passive movement can occur. The pelvis is subjected to continual minor stresses in walking, running and jumping, and occasional compression may occur in fighting or falling. If it were a completely rigid structure, many of these stresses would be transferred directly to the joints between the vertebrae, and the risks of fractures of the pelvis would be significant in severe compression. So the joint here has evolved able to absorb energy by deforming slightly under stress, without any significant freedom of movement.

What tissue do you know which has suitable characteristics for such a position? (Note 10.H)

Joints without movement frequently play an important part in patterns of growth.

Freely movable joints

The joints at which little or no movement occurs are places where bones are joined by fibrous tissue or cartilage in various arrangements. But these tissues are clearly inappropriate in joints such as the knee or shoulder, at which wide ranges of movement occur. In such movements, the surface of the head of the humerus, for example, moves very freely over the joint surface of the scapula, with sliding and shearing displacements of many millimetres. *What specialization is necessary to accommodate such movements? (Note 10.I)*

Such joints are often required to move while bearing considerable loads, as in the knee joint while running. Rigid joint surfaces of bone would be extremely difficult to lubricate effectively in these circumstances. A tissue with a certain amount of springiness, able to deform under stress to provide a relatively broad area of contact in any position of the joint, would obviously be an advantage here.

Suggest a tissue with which the bone surfaces could be covered at movable joints, which would have these characteristics. (Note 10.J)

So the basic design of a synovial joint is a fluid-filled sac, from which cells, extracellular fibres and blood-vessels are excluded. The surfaces of the bone taking part in the joint are covered with a thin layer of hyaline cartilage.

Synovial membrane. This is the name given to the lining that extends between the ends of the bones, surrounding the fluid-filled joint cavity. It consists of an inner layer of cells, the mesothelium, supported by a very vascular connective tissue (Fig. 10.13a). Outside this is a tough capsule of collagen, preventing the tearing of the delicate synovial membrane by excessive joint movement. In the early embryo, the layer of mesothelial cells surrounds the entire cavity, covering the cartilage at the bone ends, but this very delicate layer soon disappears from the cartilage as movements start.

The mesothelial cells of the synovial membrane have three major functions. The first is the synthesis of GAGs present in the synovial fluid. The second is the removal of debris and particulate matter from the joint fluid. These two functions are the properties of two rather different cell types, which together make up the mesothelial layer.

Describe, briefly, what you would expect these two cell types to look like (Note 10.K)

The third major function of the synovial membrane is the formation and renewal of the fluid in the joint cavity. This fluid is virtually an ultrafiltrate of plasma, in other words, extracellular fluid. The capillaries in the synovial membrane are responsible for a larger volume of extracellular fluid than capillaries in connective tissue elsewhere, and several specializations exist to assist the circulation of fluid. The capillaries are believed to be fenestrated.

What does this mean? (Note 10.L)

There is no basal lamina underlying the mesothelial cells, nor are the latter joined to each other by tight junctions. Fluid can therefore flow freely from the extracellular space around the capillaries into and out of the joint cavity.

Articular cartilage and joint fluid. Articular cartilage at the bone ends extends usually up to the attachments of the synovial membrane. It is a type of hyaline cartilage, and usually forms a layer 2·5 mm thick. The attachment to underlying bone follows an irregular line, with many blood-vessels in the adjacent bone.

Articular cartilage (Fig. 10.13b) has an elaborate organization, varying from joint to joint and from one position in the joint to another. In general, bundles of collagen fibres deep in the cartilage are arranged at right angles to the bone–cartilage junction, but these bundles, followed towards the joint surface, are arcades, rather like hairpins. In the surface layer, these bundles are reinforced by many bundles of collagen running parallel to the surface, their orientation resisting the tensile stresses produced by weight-bearing. The chondrocytes, which are typically plump and rounded in their lacunae deep in the cartilage, appear flattened near the joint surface.

Between these cells and fibres lie the proteoglycan aggregates with bound water.

Movement and synovial joints*.* Let us consider first a rapid movement, which brings into contact with the opposing joint surface an area of articular cartilage which was not previously under load.

What is the response of hyaline cartilage to a deforming force of short duration? (Note 10.M)

The result of these events is that the cartilage surface deforms, bringing a wider area of cartilage into contact with the opposing surface. The load on the cartilage is thus spread over a greater area. This obviously reduces the stress on the cartilage at any given point, and the risk of damaging the surface. At the same time, lubrication, or the reduction of friction between the two surfaces, is made easier, since friction increases with the vertical force driving the two surfaces together, and the deformation has reduced this force by distributing it over a wider area of contact.

Now consider the behaviour of articular cartilage under continued loading, such as occurs in the knee-joint on standing still for a long time.

How does hyaline cartilage behave under such conditions? (Note 10.N)

So prolonged loading increases still further the areas of contact, and, in addition, drives fluid out of the cartilage into the joint cavity.

Note that pressure on the cartilage cannot reduce its blood supply, since the capillaries supplying it are protected in the underlying, rigid bone, or free from compression in the synovial membrane.

Synovial fluid lubricates the opposing joint surfaces, both by a layer of GAGs adsorbed on to each surface, and by free GAGs in solution — referred to by engineers as boundary and fluid film lubrication respectively (Armstrong and Mow, 1980). The viscosity of the lubricant must vary with the vertical load on the two surfaces, to maintain effective fluid film lubrication: at high loads, a very thin lubricant would be driven out from between the opposing surfaces. The consistency of synovial fluid varies greatly in mammalian joints. It is thin and watery in the joints of the mouse, and thick, almost like butter, in the weight-bearing joints of the larger mammals.

We have so far discussed synovial joints in terms of facilitating movement. Unwanted and excessive movements must be prevented. This is to some extent the function of the tough capsule around each synovial joint.

Suggest a design for a structure that could reinforce the capsule, preventing unwanted movement without restricting other movements. (Note 10.O)

Summary

Increasingly, as you progress through this book, you are able to use familiar properties of cells and tissues to predict histological patterns of organization, and to relate these to the structures and functions of anatomy and physiology. This chapter has introduced very little that is quite new. It has taken the elements of connective tissue and reassembled them in new combinations to produce most of the specialized tissues and structures of the skeleton, helping you to visualize how the contractility of muscle is converted into useful movement.

Further reading

Hancox, N.M. (1972) "The Biology of Bone." No. 1 of the series Biological Structure and Function. Cambridge University Press. A good, general review. Try the chapters on the osteoblast, osteoclast and osteocyte.

Simkiss, K. (1975). "Bone and Mineralization." Studies in Biology, No. 53. Edward Arnold, London. An unusual book, presenting several of the classical experiments into mineralization in their historical context, with discussion. Very readable.

Stockwell, R.A. (1979). "Cartilage Cells." Cambridge University Press. A good, recent review, but rather disappointing illustrations. Try the section on the pericellular environment.

The remaining references are all chapters in "Scientific Foundations of Orthopaedics and Traumatology" (1980), edited by R. Owen, J. Goodfellow and P. Bullough, and published by Heinemann, London.

Ali, S.Y. Mechanism of calcification. Chapter 23, pp. 175–184. Well written and up-to-date summary.

Armstrong, C.G. and Mow, V.C. Friction, lubrication and wear of synovial joints. Chapter 28, pp. 223–231. A clear account of a subject often poorly handled in textbooks of anatomy and histology.

Kempson, G.E. The mechanical properties of articular cartilage and bone. Chapter 8, pp. 49–57. An engineer's account, with clear definitions of terms at the beginning.

Rosenberg, L.C. Proteoglycans. Chapter 6, pp. 36–41. The clearest account I have read of the biochemical and biophysical properties of these molecules.

In addition, many other chapters may interest you, linking histological structure to anatomy and orthopaedics.

11
A look at tubes

Many of the structures out of which the body is built are tubes — the gastro-intestinal tract, the kidneys, the blood-vessels, and so on. Tubes convey air to the lungs, urine, sweat and bile to the exterior, while our reproductive tracts are two series of modified tubes. These multitudes of tubes are built out of relatively few elements and follow the same basic patterns — this is what makes them so difficult to sort out on the basis of pattern recognition alone. Looked at another way, however, the patterns in which the cells and extracellular materials are organized to make a tube tell us directly about the functions of the tube, and allow us to predict, often with surprising accuracy, where a particular section has come from. In other words, use your intelligence and your powers of deduction to extract information from histological sections, instead of relying solely on memory and pattern recognition.

Using only the information you have already had, this chapter will indicate some of the basic links between the structure of a tube and its function. Assume you are looking at a section, without knowing where it is from, and you find a tube to be present.
In what sequence do you examine a section? (Note 11.A)

General observations about the tube

Size is probably the first thing to notice. Some tubes will be obvious on naked-eye inspection of the section, others will be only a few micrometres in diameter. Now it is clear that exchange between the contents of any tube and the tissues around takes place at the edge of the lumen, so that the ratio on cross-section between the perimeter of the lumen and its area should be very high for effective and rapid exchange: in other words, exchange in more efficient in tubes of small diameter. In nearly every system of the body, tubes that are the site of considerable exchange are small, while tubes that convey material unchanged from place are larger; tubes that are sites of storage are larger still. In the vascular system, capillaries are only a few micrometres in diameter, while arteries and veins may be measured in centimetres. The tubules of the kidney in which urine is formed are similar in size to capillaries, the ureters which carry the urine from the kidneys are several millimetres in diameter (Fig. 11.1) while the urinary bladder is much larger still.

Is the tube single, or are there many similar tubes present? (Fig. 11.1). Is the tube obviously coiled (Fig. 6.4), or is it straight? A single straight tube is likely to be conveying something from one place to another without altering its composition much. Coiling and the presence of many similar tubes in parallel are both modifications to increase the interface between the surrounding tissues and whatever is inside the tube.

(a)   (b)

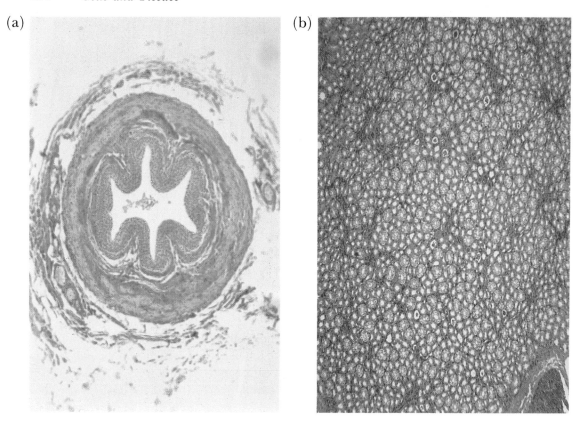

Fig. 11.1   Tubes with large and small diameters. (a)   Ureter. H & VG. (× 60). (b) Tubules in the medulla of the kidney. H & VG. (× 60.)

Be prepared to see tubes cut, not only in strict transverse section, but longitudinally or at some intermediate angle (Fig. 11.2).

Is muscle present in the wall of the tube? Tubes with muscular walls are involved in conveying their contents from one place to another: usually, they will be larger in diameter than tubes in the same system that are concerned only with exchange, and they will not often be multiple or coiled. The presence of a layer of muscle cells between the lumen and the surrounding tissues is an obvious barrier to exchange, and tubes that are specialized for exchange lack a muscle coat. The gastro-intestinal tract is unusual in this respect: it is a site of obvious and rapid exchange between lumen and body tissues, and it needs to propel the contents along. How it is designed to serve both functions will be discussed in Chapter 14.

Muscle coat present

Let us first assume that the wall of the tube contains muscle. Identify the type of muscle present.
What are the histological characteristics of smooth, striated and cardiac muscle? (Note 11.B)
If, as is usually the case, smooth muscle is recognized (Fig. 11.2), your knowledge of the

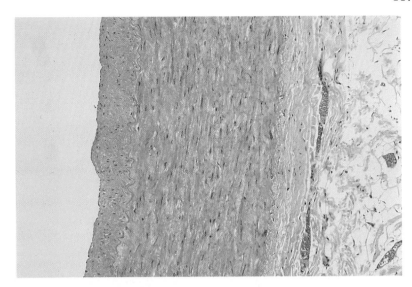

Fig. 11.2 The walls of arteries.
(a) Transverse section, lumen to the left. H & E. (× 100.)

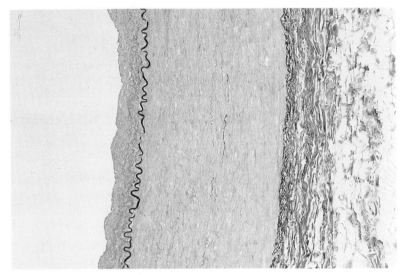

(b) Same. Verhoeff's & VG. (× 100.)

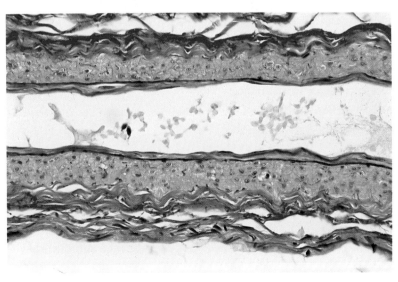

(c) Longitudinal section of smaller artery. Verhoeff's & VG. (× 400.)

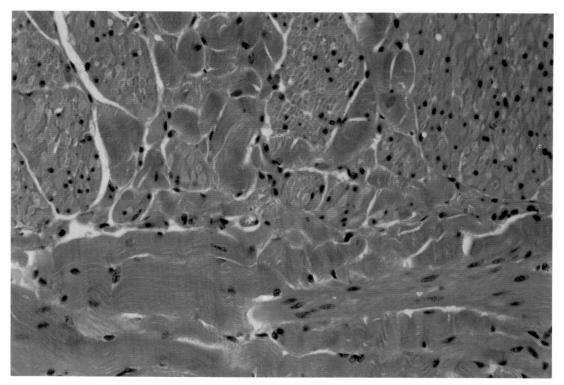

Fig. 11.3   Muscularis externa of the upper third of the oesophagus, with smooth and striated muscle in circular (below) and longitudinal layers. H & E. (× 345.)

functional characteristics of this cell type will tell you how rapidly contraction and relaxation take place. If striated muscle is present (Fig. 11.3), a different speed of contraction is occurring, while cardiac muscle has its own characteristics.

What are the functional characteristics of smooth, striated and cardiac muscle? (Note 11.C)

The arrangement of the muscle in the wall of the tube tells us about its function. If there is a single layer of muscle, arranged concentrically around the lumen — a circular layer — its contraction will make the lumen smaller. Such a layer cannot propel the contents along, but serves to regulate the flow. It is always smooth muscle, and it is seen very clearly in the walls of arteries (Fig. 11.2). To move the contents down a tube requires two or more layers of muscle fibres, arranged at right angles to each other; these layers are often an inner circular and an outer longitudinal. The gastro-intestinal tract, with its peculiar requirements for motility and exchange, has a third, narrow layer of muscle present, in between the inner circular layer and the lumen, the muscularis mucosae (Fig. 11.4).

What other components accompany muscle in the wall of the tube?

Collagen is always present to some extent, surrounding the muscle fibres. Elastic fibres always occur, though often sparsely. In some tubes, however, the relative amounts of collagen are increased: this implies that the tube will resist distension beyond a certain level. The increase in amount of elastic fibres can be even more dramatic. Figure 11.5 shows a section through the wall of the aorta: here, it is not easy to find smooth muscle fibres, and the whole thickness of the wall looks like a feltwork of collagen and elastin. Clearly, regulation of flow through the tube is not of major importance: instead, the tube is designed to withstand sudden, intermittent rises in pressure in the lumen, corresponding

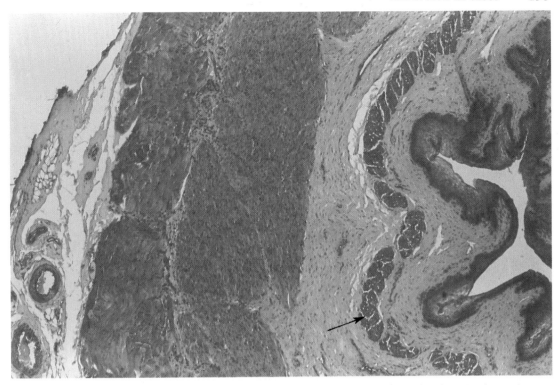

Fig. 11.4   The wall of the human oesophagus: muscularis mucosae (arrow). H & E. (× 30.)

with the contraction of the left ventricle. The pulse of pressure distends the wall, stretching the elastic fibres, until further distension is prevented by the alignment of the collagen fibres. As soon as ventricular contraction stops, the recoil of the elastic fibres returns the aorta to its original diameter. The presence of a circular layer of elastic tissue in the wall of the tube should always suggest intermittent pulses of high pressure within the tube. Arteries in which flow is regulated by a circular muscle coat also have this high content of elastic fibres, characteristically arranged in inner and outer elastic laminae, inside and outside the layer of muscle (Fig. 11.2).

Some tubes possess many elastic fibres, arranged longitudinally in irregular bundles.

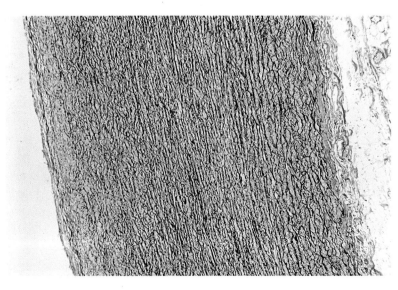

Fig. 11.5   The wall of the aorta: lumen to the left. Verhoeff's & VG. (× 60.)

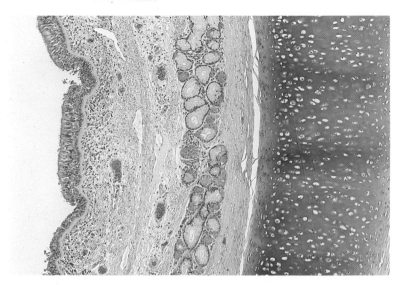

Fig. 11.6   The wall of the human trachea: lumen to the left. H & E. (× 60.)

These include the bronchi and bronchioles, carrying air to and from the alveoli of the lungs. These tubes are also subjected to intermittent forces, from which they recover by elastic recoil. As one breathes in, the volume of the thoracic cavity expands, stretching the elastic fibres whose recoil contributes significantly to the movements of breathing out.

The walls of the air passage show one further adaptation. They often contain curved bars of cartilage. Many large tubes in the body are closed most of the time, with their opposite walls in contact, only opening up as material passes down the lumen. In other words, the increase in pressure within the lumen opens the tube. In contrast the tubes that transport air to and from the lungs must remain open at all times. As one breathes in, the pressure within the lumen can be surprisingly low, yet the tube must be held open.

These tubes must have walls sufficiently rigid to keep them open, yet flexible enough to permit twisting and elongation. Hence the presence in the wall of hyaline cartilage (Fig. 11.6), arranged as C-shaped bars which are joined by tissue rich in collagen and elastic fibres, and containing a few smooth muscle fibres.

The structure of the wall of a tube that contains muscle fibres

To sum up so far, the presence of a recognizable muscle layer in the wall of a tube suggests that transport of the contents is the primary function, rather than exchange. One circular layer indicates that the tube is capable of constricting, to regulate the flow of contents, but that it is not responsible for propelling the contents. Two or more layers of muscle at right angles to each other are characteristic of tubes that do propel their contents. A third muscle layer, separate from the main muscle coat and on the side nearer the lumen, is the muscularis mucosae, and only occurs in the gastro-intestinal tract, where exchange must coexist with propulsion of contents.

A high content of elastic fibres indicates intermittent stretching and recoil of the walls of the tube: if the fibres are concentrically arranged, this stretching is distension; if the fibres are longitudinal, the stretching is along the axis of the tube.

The presence of cartilage implies that the lumen must be kept open at all times, a condition seen in the larger air tubes of the respiratory tract.

The presence of a muscle coat, whether or not it contains elastic fibres and cartilage, divides the wall of the tube into three regions: that inside the muscle, the muscle itself and

ARTERY

DUCTUS DEFERENS

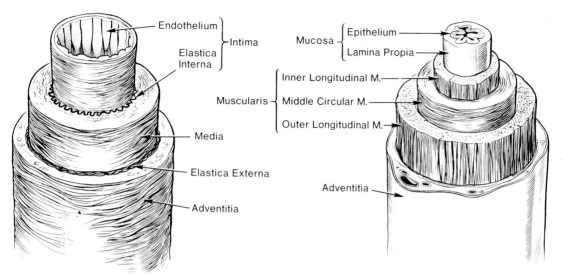

SMALL INTESTINE

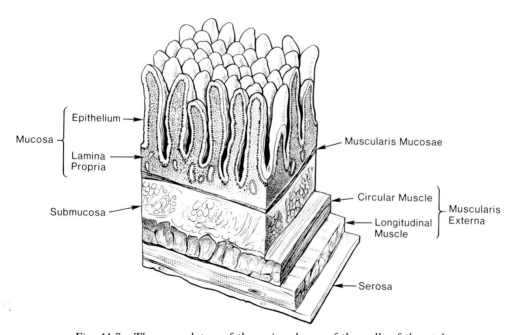

Fig. 11.7 The nomenclature of the various layers of the walls of three tubes.

that outside the muscle. The words used to describe these regions differ from one tube to another. In tubes that are lined by epithelium, the innermost region is called the mucosa, and it is made up of the epithelium itself, lining the lumen, and the loose connective tissue, rich in blood-vessels, between the epithelium and the muscle: this is called the lamina propria (Fig. 11.7). The muscle coat is the muscularis, and outside this is a region of fairly dense connective tissue, merging with the loose connective tissues of the rest of the body in a gradual fashion, with no clearly drawn line of separation. This outermost region is called the adventitia. In the the gastro-intestinal tract, the region inside the muscularis mucosae is the mucosa, composed of epithelium and lamina propria. The region between muscularis mucosa and the muscularis externa is called the submucosa. Blood-vessels are the prime example of tubes that are not lined with epithelium. In such tubes, the innermost region is the tunica intima, with endothelium lining the lumen, while the muscle coat or its equivalent in the aorta is called the tunica media. The outermost region is called the adventitia.

The region inside the muscle layer. In blood-vessels, the tunica intima consists of the endothelial layer only, in all but the very large vessels, and even in these the sub-endothelial connective tissue is sparse and thin (Fig. 11.2). Endothelial cells are thin, flattened cells when the vessel is distended, but may look plumper and even cuboidal in a collapsed vessel. In arteries, subjected to distension with each contraction of the ventricles, the circular elastic fibres are usually very obvious, forming an inner elastic lamina at the junction of intima and media, which often looks crenellated, like the battlements of a castle, as the elastic fibres shorten with the loss of pressure within the lumen. Red blood cells may be seen in the lumen, though you must be prepared to identify blood-vessels without red cells, particularly in animal material fixed by perfusing fixative through the vascular system.

In what ways would you expect the wall of a vein to differ from that of an artery of similar calibre? (Note 11.D)

Turning to tubes lined by epithelium, a lot can be deduced from looking at the mucosa. Taking the epithelium first, is it simple or multi-layered?

What deductions can one make from this observation? (Note 11.E)

The major characteristics of epithelia have been dealt with in Chapter 4. You should be able to recognize with confidence the epithelia that face urine, or that face a moist surface subjected to abrasion. Epithelia that secrete proteins intermittently, or that produce and move along a layer of mucus, that absorb materials from the lumen, or pump ions into or out of the lumen have specific structures and appearances with which you are now familiar. This information can sometimes be supplemented by looking at the lamina propria. Look for the presence of glands in this layer: if present, what types of cell are visible? Are they an extension of the cell type lining the lumen, or a different cell altogether? Use their presence to help build up a mental picture of what is happening to the material in the lumen. Is it being altered during transit by, for instance, enzymes secreted by mucosal glands, or is its passage just being helped by the secretion of lubricating mucus? (Fig. 11.6).

Blood-vessels and lymphatics in the lamina propria often form a very dense plexus, particularly where active exchange is going on across the epithelium, but these very fine vessels often collapse down in preparing the tissue for microscopy, and may not be very obvious in standard sections.

One other feature to look for is the presence of small nodules of lymphoid tissue in the lamina propria. These occur below nearly every epithelial surface, but their frequency and size vary greatly. Their frequency is quite closely correlated with the probability of invasion by micro-organisms. To take the gastro-intestinal tract as an example, the contents are reasonably effectively sterilized by the acid secretions of the stomach, and the stomach itself and the first part of the duodenum have only small and scattered nodules of lymphoid tissue. As the small intestine nears the caecum, the probability of finding bacteria which have migrated up from the huge colonies normally resident in the large intestine increases, and so do the frequency and size of nodules of lymphoid tissue. Small bronchi have many nodules of lymphoid tissue, particularly where they divide; the pancreatic duct, opening into the relatively sterile duodenum, does not.

The region outside the muscle layer. By contrast, much less information can be gained from examining the adventitia. Two points are worth noting.

The first is whether or not a serous coat is present around the tube. Loose connective tissue, particularly when full of fat cells, can accommodate a certain amount of sliding and distension of one structure relative to its surroundings. The shortening and fattening of muscles when they contract is usually accommodated in this way. But unusually large sliding movements of one structure on another may exceed the capacity of loose connective tissue to adapt. The body meets these sliding or shearing movements by interposing a closed sac of fluid, with certain characteristics.

What are the major characteristics of such fluid-filled sacs? Give four examples of such sacs. (Note 11.F)
If, when you look at the connective tissue outside the muscle layer of a tube, it ends in a smooth, natural surface on which a single layer of flattened cells may be visible, the tube faces one wall of such a sac (Fig. 11.8), and you may at once assume that large shearing movements occur between the tube and the structures on the other side of the sac. Often, movements of the tube itself are responsible. The movements of peristalsis in the small intestine, for instance, require that it be separated by a sac lubricated with fluid from the

(a)   (b)

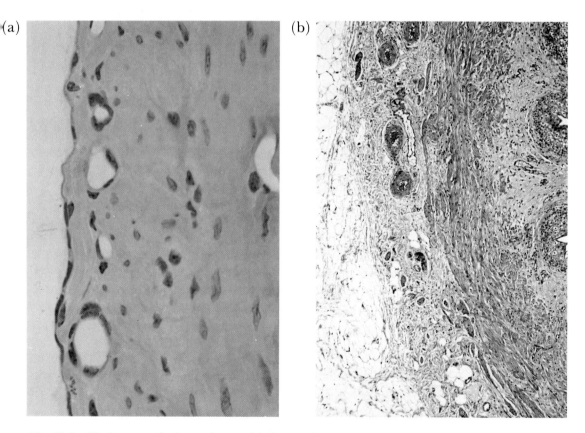

Fig. 11.8   The layer outside the muscle coat. (a) Serosa of small intestine. H & E. (× 600.) (b) Adventitia of ureter. H & E. (× 130.)

structures around it. But note that the adjacent structures, such as the posterior abdominal wall, also face into the sac, needing a serosa if the intestine is to move across them. The presence of a serous coat does not necessarily indicate that the structure itself moves significantly, but only that movement takes place between the structure and adjacent ones. The movements of the diaphragm result in organs such as the spleen moving, passively, considerable distances relative to the walls of the abdominal cavity.

Students initially find the recognition of such a layer difficult. But a block of tissue cut out of connective tissue has a much less regular surface when viewed down the microscope, however sharp the scalpel and careful the operator, and structures cut across can always been recognized (Fig. 6.1). The redundant names for such a serosal surface also cause confusion.

How many names can you find for the smooth layer on the outermost surface of a section of small intestine? (Note 11.G)

The second helpful observation that can be made from looking at the adventitia is that, in some situations in the body, the wall of the tube appears to be attached firmly to underlying bone or cartilage. This would obviously have the effect of making the wall rigid, preventing collapse.

Where might the body have tubes of such a structure, and what would occupy the lumen? (Note 11.H)

The distribution of vessels and nerves to tubes

Tubes with many-layered walls require a blood supply. It sometimes surprises students to find that this is true, even if the tube is itself an artery carrying oxygenated blood. From what we have said about the ideal relationship between the surface area of the lumen and its volume, it should be clear that a broad column of arterial blood is not as effective a source of oxygen to surrounding tissues as the capillary network into which it opens. The muscle cells of the tunica media require oxygen. This is supplied to all but the innermost layers by small vessels entering the arterial wall from the adventitia — the vasa vasorum.

The arrangement of blood-vessels has some common features in any tube with multi-layered wall. Blood-vessels approach through the connective tissue of the adventitia and run on the outside of the tube, sending radial branches towards the muscle coat and the intima or mucosa, as the case may be. In a transverse section through such a tube, the largest vessels will usually be found in the adventitia, and recognizable vessels will become smaller and smaller as one scans in towards the lumen.

Similarly, the autonomic nerves exerting central control over the activity of the muscle coat approach through the adventitia, often in company with the blood-vessels. These nerves end in a local plexus, which in the gut contains intrinsic neurones, whose activity they influence. In the gastro-intestinal tract, these intrinsic neurones are clearly visible, forming the myenteric plexus of Auerbach between the two layers of the muscularis externa, and the submucous plexus just external to the muscularis mucosae (Fig. 11.9)

In tubes with a serous layer on the outside, it is clear that the nerves and vessels can only reach the tube by travelling between the layers of serous membrane that connect tube to body wall — in the case of the intestine, the mesentery.

Tubes modified to form storage chambers

Tubes are often modified to form storage chambers: the bladder, the gall-bladder and the

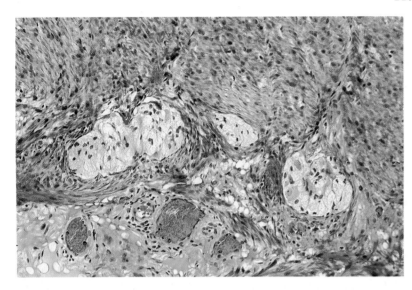

Fig. 11.9 The myenteric (Auerbach's) plexus of the colon. Circular muscle, below: pale areas at centre are bundles of nerve fibres and neurones. Masson's trichrome. (× 150.)

stomach are three examples. The modifications to the structure of the walls are often surprisingly minor. Usually, the circular muscle coat is thick and prominent at the outlet of the tube, providing a sphincter, and the lumen is correspondingly larger than elsewhere in the tube. If changes in volume are substantial between the full and empty states, it is likely that the storage chamber will face a serous sac on at least one surface.

As the storage chamber fills and pressure rises inside it, contents might be expected to back up the tube that enters it, dilating it progressively. In fact this does not normally happen. One common device for preventing it is for the tube or tubes to enter the storage chamber obliquely, with a relatively long course through its wall. As the chamber expands, the terminal part of the tube is compressed where it travels through the wall, reducing the chance of reflux. The entry of the ureters into the bladder is an example of this mechanism.

Tubes lacking a muscle coat

Such tubes are specialized for exchange between the contents and the extracellular space. They tend to have a small diameter, with a consequently high surface area compared to their volume. The wall is reduced to the minimum possible thickness, to assist the processes of exchange, so there cannot be the structural variety seen in larger tubes with many-layered walls. Capillaries and lymphatic capillaries are prime examples of tubes designed for exchange (Fig. 7.7). The flattened endothelial cells are separated from the extracellular space only by a basal lamina. The formation of urine in the kidneys requires a great deal of exchange between the kidney tubules and capillaries in the extracellular space, with different portions of tubules often influencing each other across the space. The kidney medulla, in particular, is built up of hairpin loops of tubules accompanied by similar capillary loops, producing a complex pattern of parallel tubes (Fig. 11.1b).

The alveoli of glands are another example of tubules designed for exchange between lumen and tissue spaces. In sorting out the functions of all these exchanging tubules, one can only note the characteristics of the cells lining the lumen, and the relationship of the tubule to other structures.

Structure and function in tubes

This chapter has dealt with a wide range of histological structures, providing links between organization and function which will help you to deduce the function of tubes you have not seen before and to predict the appearances of tubes the functions of which are known. Let us end it by trying both types of deduction.

The human pancreas is a large gland that produces digestive enzymes, including a number that are proteolytic. This secretion is intermittent, and is carried by the pancreatic duct system into the second part of the duodenum to coincide with the arrival of food from the stomach. The transport of fluid containing these enzymes, in the form of inactive precursors, is achieved by fluid pressure in the acini, which sufficiently high during active secretion to ensure flow down the ducts without need for muscular activity.

The contents of the stomach are very acid, yet the pancreatic enzymes all have pH optima near to neutrality, so the fluid carrying pancreatic secretion into the duodenum must neutralize the acidity of the gut contents. To achieve this, there is a large volume of pancreatic fluid, and it is rich in sodium bicarbonate. The cells lining the duct system add water to the secretions, and also add bicarbonate ions.

Now let us design a tube for these functions. It does not propel the contents, and regulation of flow is achieved in the acini, not along the course of the duct.
Will the wall contain muscle fibres? If not, what is likely to be its structure? (Note 11.I)
These ducts will be larger than capillaries since their major function is transport. They do modify the contents, however, requiring exchange between contents and the tissues around the duct.
Will there be a simple epithelium, or a compound one? Are there likely to be capillaries — vascular or lymphatic — closely associated with the tube? (Note 11.J)
The epithelium lining the duct is passing water and bicarbonate ions into the lumen, but not pumping any ion against a powerful gradient.
What type of cell would you expect to see in the epithelium? (Note 11.K)
Finally, a few additional features can be predicted. The ducts undergo no great changes in shape or position relative to their surroundings. They open into a practically sterile part of the gut by a narrow opening, and are further protected against bacterial invasions by the flow of their contents into the duodenum.
Will this tube have a serosa or a typical adventitia? Would you expect to see mucosal aggregates of lymphoid tissue? Is the mitotic rate in this epithelium likely to be high? (Note 11.L)
The light microscopic structure of pancreatic duct is shown in Fig. 11.10.

Now let us do the trick the other way round.

Figure 11.11 is a section through a tube. What can we deduce about its function from its appearance?
First, its size is much larger than the diameter of a red blood cell; cell nuclei in its lining layer number hundreds in a single cross-section. It exists for transport rather than exchange.

That being so, does it have a muscle layer? It certainly does, with smooth muscle arranged in an inner circular layer and an outer longitudinal; there is even a suggestion of scattered longitudinal fibres inside the circular layer. With such a muscle coat, the tube clearly transports its contents by peristalsis.

Turning to the mucosa, the epithelium is simple, with tall, columnar cells carrying long apical projections which look like cilia; there are none of the mucous goblet cells associated with ciliated epithelia in the respiratory tract. In fact, if we were to examine these

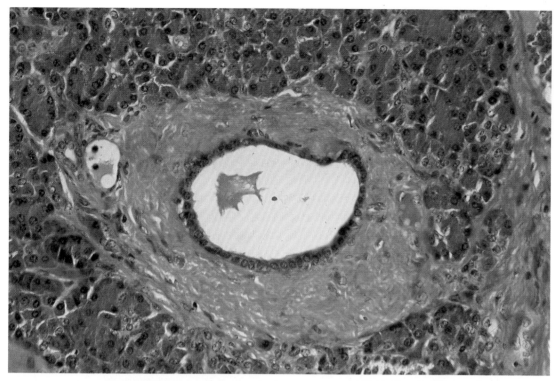

Fig. 11.10   A small duct in the human pancreas. H & E. (× 345.)

(a) 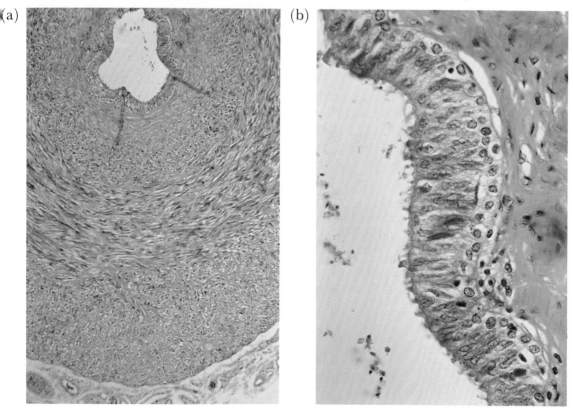 (b)

Fig. 11.11   Transverse section through a tube. Masson's trichrome. (a) Survey view (× 60.) Two small wrinkles in section near lumen. (b) Detail of mucosa. (× 385.)

projections in the electron microscope, they lack the characteristic internal structure of cilia. They are stereocilia, or massive microvilli.

What is the characteristic structure of a cilium? What functions do you associate with microvilli? (Note 11.M)

Between the bases of the columnar cells, many small, darkly stained nuclei can be seen, belonging to cells resting on the basal lamina. The epithelium surrounds a small, irregular lumen. Putting these features together, the tube is closed, but capable of distension. Some exchange takes place with material in the lumen, mainly absorption, with no clear evidence for secretion.

The lamina propria is thin, with some collagen and elastic fibres, but no glands. Once again, it does not look as if significant amounts of material are being added to the luminal contents. Little lymphoid tissue is present, so the chance of infection is low.

Turning to the adventitia, it merges gradually with connective tissue, and there is no serous membrane. The tube does not move much or distend very greatly, nor do the structures adjacent to it move much.

The section is actually of the ductus deferens, the very muscular tube that transports sperm from the epididymis to the urethra, a rapid and intermittent process in which nothing significant is added to the luminal contents. Following ejaculation, however, residual sperm and the fluid in which they are carried are resorbed from the lumen.

The cellular organization of the body follows reasonable, logical rules. With intelligence and a little information, you can often predict the unseen, and interpret the functions of structures seen for the first time.

Communication systems

Patterns imply organization: organization is impossible without communication, or the transfer of information. The patterns we recognize involve the size, shape, spacing, orientation and activity of many thousands of cells, each behaving in a reproducible and disciplined way that is appropriate to its position in the tissue or organ concerned. This reproducible behaviour is partly due to local information reaching the cell from many surrounding cells and cell groups, and from the extracellular materials — collagen, basal laminae, and so on — produced by cells. It is also influenced by cells in distant parts of the body.

Historically, scientists discover the dramatic features of living organisms first, and concentrate their attention on them. So the rapid and precise transfer of information over long distances by nerves has been intensively studied. By contrast, the local messages that allow one cell to influence its neighbours remain almost unknown. Yet these local communications are largely responsible for structural homeostasis, maintaining the detailed organization of cells into their functional patterns. These local messages are probably much older in evolutionary terms than nerve conduction.

The simplest way in which one cell can influence another is the release of some signalling molecule into the extracellular fluid, followed by its diffusion across to the recipient cell. Further possibilities for communication exist if the two cells make contact. Both of these methods of communication are only possible if the cells are close to each other. Two specializations of the former method, diffusion, have developed to allow cells to influence events at distant sites. The first is signalling by hormones, which is essentially the release of signalling molecules into the blood stream for rapid distribution around the body; the other is nerve conduction, in which a cell reaches out a very long, thin process to release diffusible molecules very close to the recipient cell. The nerve impulse is the device by which the nerve cell body controls the release of signals from the tips of processes a long way away.

Local communications between cells

Evidence for the importance of local messages in vertebrates comes mainly from experimental embryology. In the course of development, signals produced by particular cell groups at specific times are crucial in determining the subsequent fate and orientation of neighbouring structures. In the adult, we know that damaged tissue produces signals which attract polymorphonuclear leucocytes and monocytes and alter the patterns of blood

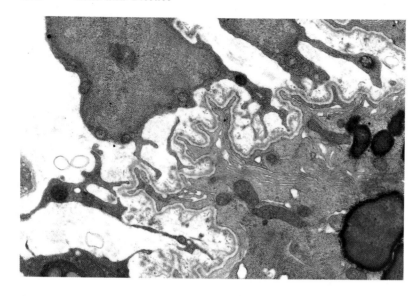

Fig. 12.1 TEM of epithelial cells (right) and underlying connective tissue cells (left), separated by a basal lamina: from rat uterus. (× 6000.)

flow locally. We have many hints of other local messages producing order and pattern in cell populations. Technical difficulties, however, have slowed down the detailed study of these communication systems.

Local diffusion of signals

It seems clear that locally diffusing messages carry information between cells and cell groups, and we can make a number of intelligent guesses about their release and detection. We would expect such signalling molecules, if they are proteins or peptides, to be membrane-bound, in vesicles, in their cell of origin, and to be released into the extracellular space by fusion of the vesicles containing them with the cell membrane. Diffusion is a relatively slow and inefficient method of transport, and the concentration of signalling molecule would fall off rapidly with increasing distance. We would therefore expect recipient cells to be very close to the signalling cell, and to have receptor sites for the signal on the cell membrane. Figure 12.1 shows the sort of structural specialization that is associated with the transfer of information by local diffusion (Cunha, 1976; Lehtonen, 1976), in this case between connective tissue cells and the epithelium lining the lumen of the rat uterus (Wischik and Rogers, in press). At the light microscope level, however, there is little to indicate this type of communication in action, though we know it must be there.

Cell contacts

Some cell types, such as fibroblasts, grow very readily in tissue culture, and migrate freely over the bottom of the culture dish when they are widely separated from each other. If fibroblasts make contact, however, they cease to migrate. This contact inhibition of movement is just one example of an influence on cell behaviour resulting from cell membranes coming together. How contact produces these results is less certain. Cells can be quite particular about the cells with which they will form contacts in tissue culture, ignoring some and preferring others. It may be that some correspondence between the glycocalyx of each cell is necessary to allow membranes to come close enough to make contact.

What is the glycocalyx, and what are its characteristics? (Note 12.A)

Once the membranes of adjacent cells have made contact, specialized structures can develop with particular functions, such as desmosomes, tight junctions and gap junctions. Even desmosomes and tight junctions influence cell shape and hence behaviour. More obvious transfer of information occurs between cells at gap or low resistance junctions. *Give two examples of groups of cells linked by gap junctions. (Note 12.B)*

In considering the parts played in tissue organization by local diffusion and cell contacts, we must also look at factors which are likely to limit the spread of information by these routes. Figure 12.1 shows rather well the basal lamina intervening between epithelial cells and those in the connective tissues below, preventing the establishment of direct contact. Diffusion of some molecules can also be limited by glycosaminoglycans in the extracellular spaces, since they carry a surface charge.

Are glycosaminoglycans present in basal laminae? (Note 12.C)

Assisted diffusion of signalling molecules — hormones

Hormones are signals released into the blood stream. The flow of blood carries them rapidly throughout the body, whereas diffusion through extracellular fluid alone would be a most inefficient and slow way of transporting them to distant sites. Hormones are secreted initially into the extracellular space around the synthesizing cell, diffusing from there into the local capillaries or sinusoids. Throughout the body hormone molecules leave the blood to enter the extracellular spaces, establishing an equilibrium between blood and tissue fluid. It is clear that the signal or hormone becomes necessarily diluted in a very large volume of blood and extracellular fluid, and will reach a great many cells which do not respond to it, in addition to those that do.

The economics of hormonal signalling

A number of specializations has evolved to make this system of communication much more economical than might at first appear. In the blood, carrier proteins, mainly globulins, bind hormones in a reversible reaction, with the equilibrium point greatly in favour of binding. Now the level of free, unbound hormone in the blood determines the diffusion rates into extracellular fluid and the equilibrium levels there. Through binding, much of the hormone is retained in the blood, and protected from wasteful dilution in the tissues. Target cells possess receptors which are also proteins which bind the hormone reversibly, with even higher affinities for the hormone than the carrier proteins of the blood. Binding of hormone to cell receptors reduces the concentration in the extracellular fluid, thus increasing the diffusion gradient between blood and ECF. Loss of unbound hormone by diffusion from the blood alters the equilibrium there, reducing the total amount bound to carrier proteins. In this way, considerable amounts of hormone can be delivered to target tissues, even though the levels of free hormone in blood and extracellular fluid remain very low.

In some cases, portal venous systems help to protect the hormone from excessive dilution before reaching its target. The hormones produced in the hypothalamus, for instance, enter capillaries which unite to form small veins running into the adenohypophysis, or anterior lobe of the pituitary gland. Here, the veins empty into a second capillary bed, delivering the hormones directly to their target cells without the dilution inevitable in distribution throughout the body.

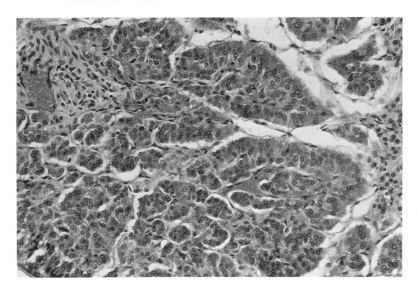

Fig. 12.2 An endocrine gland, the adrenal medulla, showing many large blood sinusoids. H & E. (× 200.)

Cells that produce hormones

Hormones fall into two main classes: those which are proteins (including glycoproteins and peptides), and those which are lipid-soluble steroids. The latter pass through phospholipid membranes readily, the former do not. This physico-chemical difference affects both the signalling cell and its target. Cells that synthesize proteins for intermittent secretion have a structure that is by now well known to you; this structure is similar whether the material secreted is glycoprotein or peptide in nature

What ultrastructural patterns do you associate with cells that synthesize and secrete steroid hormones? (Note 12.D)

All endocrine glands have a very rich blood supply (Fig. 12.2). In the majority of them, the hormone secreting cells are arranged in irregular cords, with a wide exposure of their surface membrane to the endothelium. The cells, which look rather epithelial, have no duct system and are not clearly polarized (Fig. 12.3). The blood-vessels surrounding the cells are often large, baggy sinusoids rather than simple capillaries. The slow flow through the sinusoids and the wide exposure of cell membrane to blood-vessel are both specializations to assist the diffusion of hormone molecules into the blood from the small volume of extracellular fluid around the secreting cells. One other specialization is not visible with the light microscope, but is important: the capillary or sinusoidal endothelium is fenestrated in endocrine glands.

What does fenestration mean, and what is its functional significance? (Note 12.E)

Target cells and their responses

In the target cell, receptors for the protein group of hormones will be on the cell membrane. Receptors for steroid hormones, which pass easily through membranes, are found in the cytoplasm. The arrival of a protein hormone on the external surface of the cell membrane and its attachment to a specific receptor produce effects in the cell through a "second messenger". On the cytoplasmic face of the membrane, the rate of synthesis of a molecular species, usually cyclic AMP, is altered, and this affects the functioning of the cell. In the case of steroid hormones, the hormone−receptor complex can migrate into the

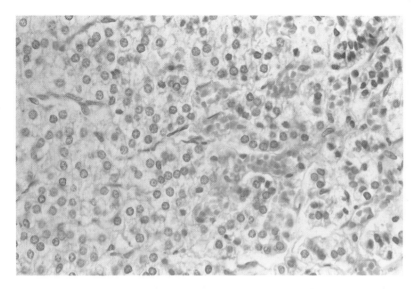

Fig. 12.3 The adrenal cortex, showing irregular cords of cells between blood sinusoids. H & E. (× 405.)

nucleus and influence gene transcription. In both systems, the arrival of a few molecules of hormone produces a sequence of events in the target cell with considerable amplification, many molecules of cyclic AMP or messenger RNA appearing for each receptor site occupied.

Note that different target cells may use the same signal as the starting point for entirely different responses. Cells in the breast, the uterus, the vagina and the brain all respond to oestrogen, but in quite different ways. Cells in each of these organs are receiving a wide variety of local messages, and the precise response of a given cell to an hormonal stimulus can be quite different if the local messages are directing it towards different behaviour patterns. The observations that the same hormonal stimulus produces responses that vary from cell to cell, and vary in the same cell at different times, can only be understood if we take account of all the information received, including local messages.

The characteristics of hormonal signalling

Hormonal signalling is in general a slow affair, producing effects minutes to hours after the first release of hormone. By contrast, stimulation of a nerve produces a twitch in the muscle it supplies in a fraction of a second. Since the signal is widely distributed in the body, a hormone can influence many groups of cells simultaneously: changes occurring in the body at puberty are the results of hormonal signals affecting growth, the development of breasts and hair, the maturation of the gonads and the genital tracts, and behaviour. On a shorter time base, the changes in the female body through the menstrual or reproductive cycle are hormonally signalled. In both cases, one hormone can influence the synthesis and release of other hormones, and also the sensitivity of target cells, so that complex patterns of synthesis and response involving a number of hormones and tissues are built up. On a still shorter time base, the release of noradrenaline from the adrenal medulla (Fig. 12.2) when one is frightened affects the cardiac output, causes dilation in some blood-vessels and constriction in others, and produces staring eyes.

In short, hormones can co-ordinate the activities of cell groups scattered throughout the body, enabling the signalling cells to produce complicated sequences of responses extending over relatively long periods of time.

Signalling by nerves

Whereas endocrine cells provide their signals to nearly all the cells in the body, relying on target cells to respond, nerve cells show great precision, liberating their signals into tiny, protected pockets of extracellular space facing the membranes of target cells. High concentrations of signal and rapid diffusion are ensured by reducing the volume of this pocket, and the distance between cells. To deliver a message so precisely to a site far from the nerve cell's body, a long, slender cylinder of cytoplasm, the nerve axon, is extended from the cell body right up to the target cell. The nerve cell instructs the terminal end of this process to release its signalling molecules, or transmitter, by means of an electrial event, the action potential, which travels rapidly down the axon from cell body to terminal.

Sensory neurones act as transducers, converting some factor in their surroundings, such as heat or pressure, into action potentials, releasing transmitters from their terminals onto other neurones. Such sites of transmitter release between neurones are called synapses. The great majority of neurones are interneurones, receiving stimulation from neurones at synapses, and integrating these into action potentials to synapses on other neurones. Motor neurones integrate the information coming to them at synapses with other neurones, and in turn release their transmitters on to muscle fibres, at neuromuscular junctions. Whether a neurone is motor, sensory or an interneurone, it acts as a minute computer, integrating the total information coming to it into a pattern of action potentials and signals to its receptive field.

We shall look first at the membrane of the nerve cell, which is concerned with the transmission of the action potential, the release of transmitters and the receipt of information at synapses. Next we shall examine the nerve cell and the structures needed to maintain its functions. Finally, we will look at the organization of nervous tissue and the ways in which other cells and structures contribute to it.

The cell membrane of the neurone

This has many features in common with the membranes of all other cells. It is a phospholipid bilayer in which a number of proteins float, some of them extending through the full thickness of the membrane. The lipid core of the membrane is impermeable to water, ions and small molecules which are not lipid soluble, but some of the intramembranous proteins provide channels through which water and some ions can diffuse. In particular, channels for the diffusion of sodium and potassium ions exist, potassium diffusing more freely than sodium. Other intramembranous proteins can actively transport ions and small molecules through the membrane: the notable example is Na−K ATPase, an enzyme which transports sodium ions out of the cell and potassium in, using energy from the hydrolysis of ATP.

The resting potential. The cell membrane separates two solutions with very different characteristics. Inside the cell are many proteins and other molecules in solution, which are largely absent from the extracellular fluid.
What mechanisms do you know for keeping the protein content of the extracellular fluid low? Why is this important? (Note 12.F)
On balance, the intracellular proteins are negatively charged. If one examines the distribution of ions on either side of the membrane, several factors are acting. Each ionic species, such as the chloride ion, diffuses through the channels available to it in response to concentration gradients, seeking to reach the same concentration on each side of the

membrane. The total electrical charge on each side of the membrane drives ions through the same channels in an attempt to equalize the concentration of positive or negative ions in cytoplasm and in extracellular fluid. Finally, the total osmolarity on each side of the membrane is equal: if it were not so, water would pass through diffusion channels by osmosis until it diluted the more concentrated solution to equal osmolarity. These three sets of forces act simultaneously. Given the intracellular concentration of negatively charged proteins too large to diffuse through channels in the membrane, and the continual extrusion of sodium ions from the cytoplasm, the final concentrations of ions produce a potential difference between the two sides of the membrane, with an excess of positive charges outside and of negative ones inside. The resting potential of the nerve cell membrane is about 75 millivolts, with the inner surface negative relative to the outer surface: this is usually written -75 mV. If the membrane potential is reduced at one point by, for instance, injecting positive ions into the cytoplasm, the effect spreads only a short way along the membrane, falling off with distance and also with time, as the distribution of ions readjusts to restore the resting potential. The lipid bilayer thus acts as a capacitor, maintaining a potential difference between its two surfaces.

Excitable membranes and the action protential. The membranes of all cells carry a resting potential. Nerve cell membranes, however, are "excitable", as are the sarcolemma and T-tubules of striated and cardiac muscles. The simplest model for this is to visualize a further set of channels through the membrane, capable of transmitting sodium ions, and guarded by two "gate" molecules (Fig. 12.4). The first type of "gate" keeps the channel

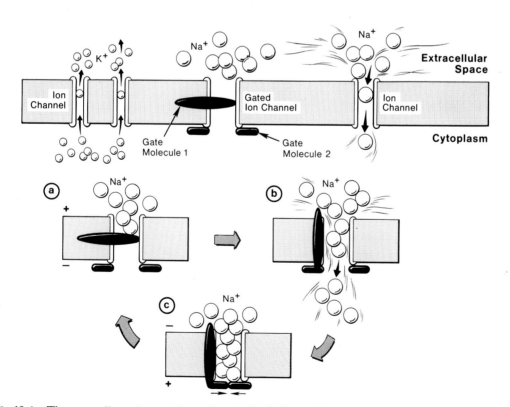

Fig 12.4   The nerve cell membrane. Above, channels for diffusion of sodium and potassium, and a gated sodium channel. (a)   Gated channel closed at normal, resting potential. (b)   Channel open at reduced membrane potential. (c)   Inner gate closed with reversal of membrane charge.

closed at the normal, resting potential, but when the potential difference across the membrane drops to a given figure — the threshold for stimulation — it opens abruptly, allowing sodium ions to pour into the cell. When the net charge on the inside of the membrane reaches a given positive value as a consequence of this influx of sodium ions, the second "gate" molecule closes the channel on its cytoplasmic side. The distributions of other ions across the membrane are affected by this sudden inrush of sodium: in particular, potassium is driven out through its diffusion channels until a new balance is reached. The potassium channels also appear to have voltage-sensitive gate molecules, which open to permit this outrush of potassium.

This sequence of events is called an action potential. Note that although it requires energy to operate the sodium pump, the action potential itself is an automatic response to voltage changes across the membrane. The opening and closing of sodium and potassium channels needs no additional energy. Note also that these exchanges of ions which underlie the action potential take place in a very narrow layer of water, no more than perhaps 20 nm thick, on each side of the phospholipid bilayer.

The entry of sodium ions in the initial stages of the action potential produces a high concentration immediately beneath the open ion channel. Some of these ions diffuse laterally on the cytoplasmic side of the membrane, reducing the resting potential. If the next gated ion channel is near enough, this drop in potential may reach the threshold for stimulation and this channel, too, will then open. In this way, an action potential can travel along the membrane of a nerve cell. The spread of an impulse over the surface of a nerve cell travels with equal ease in any direction: an axon can transmit an action potential in either direction.

Nerve impulses

The passage of an action potential down a nerve axon is a nerve impulse, which triggers the release of transmitter at its terminal.

The speed of passage of a nerve impulse down an axon is of the order of $0.5-1.0$ m/s. for small, unmyelinated nerves. Increasing axon diameter and the presence of a myelin sheath both increase conduction speeds, up to 130 m/s. in large axons with myelin sheaths.

Non-myelinated axons. Small axons without myelin sheaths do not lie free in the extracellular fluid, but are surrounded by the cytoplasm of supporting cells called Schwann cells, with a narrow, fluid-filled gap intervening between the membranes of axon and Schwann cell (Figs 12.5, 12.6). Nerve impulses in the axon do not produce action potentials in the Schwann cell membrane. Since the ionic changes associated with action potentials take place in a very thin layer of water on each surface of the nerve membrane, the gap acts as an insulator; in addition, the Schwann cell membrane is not excitable. *What is missing from this membrane that is present in the excitable membrane of the nerve cell? (Note 12.G)*

Often, several non-myelinated axons share the same Schwann cell. A basal lamina surrounds the outer surface of the Schwann cell, separating it from the general extracellular fluid.

Myelinated axons. The axon of a myelinated nerve is surrounded by a linear succession of supporting or Schwann cells, each of which wraps around a length of the axon. Once

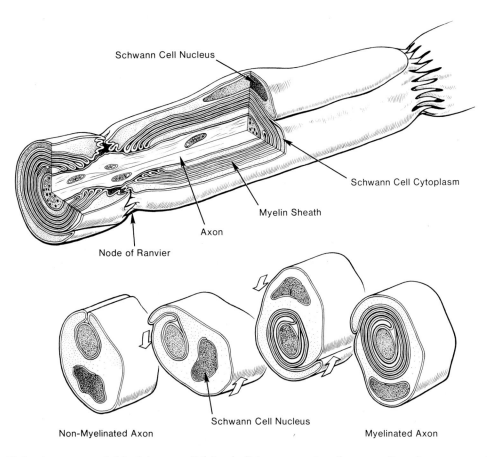

Fig. 12.5   Axon surrounded by Schwann cell (above). Below, conversion of an unmyelinated axon to a myelinated, by Schwann cell rotation to produce a myelin sheath.

clasping the axon, the Schwann cell rotates many times around it, leaving a double layer of cell membrane behind with each rotation until the axon is sheathed in multiple layers of membrane, closely applied to each other (Figs 12.5, 12.6); the cytoplasm and nucleus of the Schwann cell appear as a smooth bulge on the outer side of this sheath. Each Schwann cell and its segment of sheath meets another similar cell above and below it along the axon, with a small gap in between adjacent Schwann cells, called the node of Ranvier. At the nodes, and only there, the surface membrane of the axon is exposed to the extracellular fluid from which it is separated only by a basal lamina.

Now the excitability of neuronal membrane varies from place to place, presumably reflecting the density of gated sodium channels. It is very high at nodes, and low between them. If a myelinated axon is stimulated at a node, the action potential produced there does not appear to spread along the sheathed surface of the axon, but ion changes inside the axon and in the extracellular fluid outside the Schwann cell are sufficient to affect the next node, causing an action potential there. In effect, the nerve impulse jumps from one node to the next, instead of travelling steadily down the membrane — hence the term, saltatory conduction. This is a much faster method of conduction.

(a)

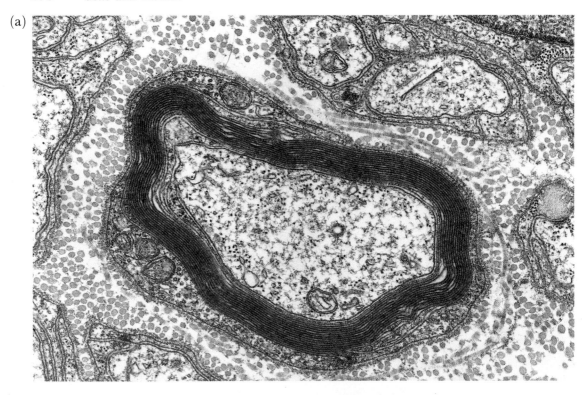

(b)

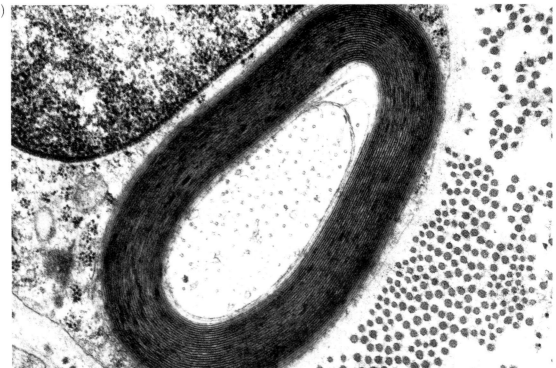

Fig. 12.6   TEMs of myelinated and unmyelinated axons. (a)   Myelinated axon (centre): two unmyelinated axons in a process of Schwann cell cytoplasm (above centre). Primary osmium fixation, giving clear staining of membranes and basal laminae. (× 37 000.) (Courtesy, Dr J. Heath.) (b)   Single myelinated axon (centre); Schwann cell nucleus (top left). Primary glutaraldehyde fixation, giving good cytoplasmic preservation: note microtubules in axon. (× 37 000.) (Courtesy, Dr M. Haynes.)

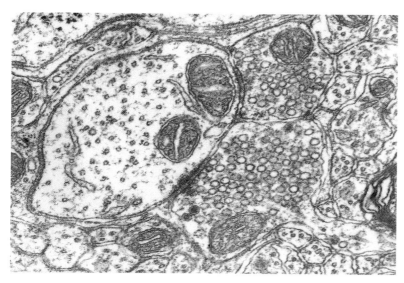

Fig. 12.7 TEM of two synapses. Neuronal process (centre of left) with microtubules and two mitochondria in contact with two axonal profiles (right of centre), filled with synaptic vesicles. (× 54 000.) (Courtesy, Dr J. Heath.)

Release of transmitter at a nerve terminal

This is the neurone's method of signalling to the receptor cell. At a nerve terminal, whether it is a synapse (Fig. 12.7) or a neuromuscular junction, electron microscopy shows a number of small, membrane-bound vesicles, which are believed to contain chemical transmitters. Occasionally one of these touches and fuses with the surface membrane, liberating its contents outside the cell. The arrival of an action potential, however, causes many vesicles to empty simultaneously. Whether we look at a synapse or a neuromuscular junction, the membrane of the receptor cell is very close, and diffusion of transmitter across the cleft takes place rapidly. It is often possible to demonstrate, histochemically, stored transmitter in nerve endings (Fig. 12.8).

Transmitters can be excitatory or inhibitory: in other words, they can reduce the potential difference at the membrane of the recipient cell, or they can cause hyperpolariza-

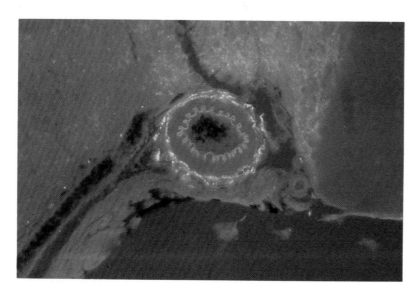

Fig. 12.8 Fluorescence micrograph of an artery (centre) on the surface of cat brain: bright, yellow-green fluorescence indicates catecholamine-containing nerves in arterial wall. (× 100.) (Courtesy, Dr J. Furness.)

tion, increasing the transmembrane potential. It seems that the nature of the receptor complex on the recipient cell determines which effect it will have. Acetylcholine is the transmitter liberated at the majority of neuromuscular junctions involving striated muscle, and at many synapses. The membrane of the recipient cell that faces the cleft is rich in receptor molecules for acetylcholine, and the arrival of transmitter molecules at these receptors permits sodium to enter the cell, i.e. it is excitatory. If enough transmitter molecules arrive, the potential falls to the threshold value, and an action potential follows in the recipient cell.

Clearly, for precise and repetitive function, all the transmitter emptied into the cleft by the arrival of one action potential at the nerve terminal must be removed very rapidly, leaving the synapse or junction free to transmit the next impulse. Various mechanisms exist to do this. In the case of acetylcholine, the enzyme acetylcholinesterase is present on the pre-synaptic and post-synaptic membranes (Fig. 2.10). This splits acetylcholine into fragments which are then taken up by the nerve terminal for recycling.

Note that synapses are uni-directional: they determine the patterns and directions of information flow in nervous tissue.

Electrical events at the nerve cell body

Neurones may have many synapses on their cell bodies, and, in addition, they may send out many processes, called dendrites, providing further surfaces for other neurones to synapse with. Figure 12.9 illustrates the dense, branching pattern of one cell's dendritic tree, which may have thousands of synapses on it. It is quite common to find a single axon forming several synapses close together on the same dendrite. How does such a complex system function?

First, the membrane of dendrites and neuronal body appears to be relatively insensitive, by contrast with that over the axon hillock, the site where the axon leaves the cell body. It is here that depolarization produces a nerve impulse down the axon. An excitatory action potential reaching an area of post-synaptic membrane on a nerve cell — whether on dendrite or on cell body — produces a local change in the membrane potential less than

Fig. 12.9 Drawing of the dendritic tree of a neurone from the cerebellum.

that needed to reach the threshold for an action potential. The simultaneous arrival of impulses at many synapses, or repeated firing of a few synapses, may result in the threshold for excitation being reached at the axon hillock, with the production of an action potential. Identical patterns of excitation at the same post-synaptic membranes of synapses may, on the other hand, fail to produce an action potential if there is simultaneous activity at inhibitory synapses. The neurone is a small computer, integrating the total pattern of synaptic activity at its surface and producing an action potential to its own terminals whenever that activity exceeds a certain threshold. These single units, nerve cells, can be built into an infinite number of interacting circuits, for the rapid yet precise passage of information between cells.

The structure of a nerve cell

If one watches a living axon by phase contrast microscopy, it is incredibly busy, with many organelles travelling down towards the terminal from the cell body, and some moving in the opposite direction. In addition, there is a slower, inexorable flow of cytoplasm down the axon, at 1–2 mm/day, that cannot be appreciated by direct observation, but becomes obvious if one labels newly synthesized proteins with a radioactive aminoacid, and looks at their distribution at later time intervals. Such labelling shows protein synthesis in the cell body, around the nucleus, with subsequent transport down the axon.

The cell body

It is clear that most of the synthetic activity needed to support the axon and dendrites takes place in the cell body. Although an axon is a narrow cylinder of cytoplasm, its total volume can be great compared to the sizes of most cells, since its length may be as much as 1 m in the human body. The size of the neuronal cell body is related to the total volume of axon and dendrites, and, in general, is large compared to other cells (Fig. 12.10). By light

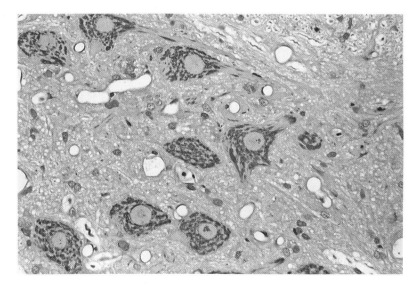

Fig. 12.10 Large neurones filled with Nissl substance in the ventral horn of the spinal cord. Cresyl fast violet. (× 385.)

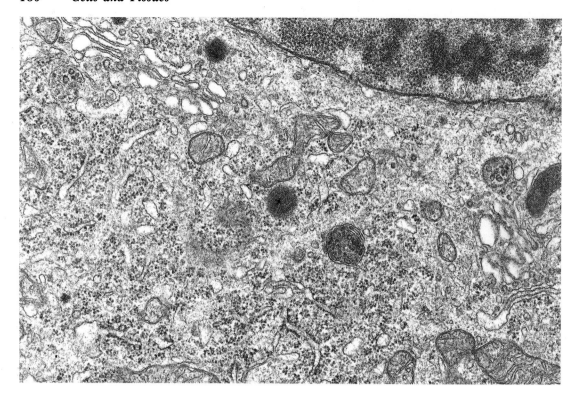

Fig. 12.11   TEM of part of cell body of a neurone in a sympathetic ganglion. (× 42 000.) (Courtesy, Dr J. Heath.)

microscopy, the cytoplasm contains granular masses of basophilic material, called Nissl substance. The TEM shows these areas to be full of clusters of free ribosomes and short runs of RER, together with mitochondria (Fig. 12.11).

The axon

Some transmitter material may be transported down the axon; in many neurones, more transmitter is synthesized in the nerve terminal itself, so that the material transported down the axon includes mitochondria and the enzymes needed for synthesis. In addition to neurotransmitters, nerve cells are known to secrete substances which have a direct effect on the recipient cell, called trophic factors. These probably play a part in establishing and maintaining the contacts between pre-synaptic and post-synaptic cell, or nerve and muscle, and evoking the development of membrane receptors for the appropriate transmitter. The retrograde transport up the axon appears to inform the neurone of the nature of the cell at the terminal. It seems that neurones compete for space on membranes of other neurones, with a continual replacement of synapses. The astonishing specificity of the connections that are formed in the brain requires information flow from terminal to cell body if appropriate synapses are to be retained and inappropriate ones allowed to die off.

Transport of organelles within cells appears to be guided by microtubules and neurones are no exception. Axons contain such tubules, running longitudinally from cell body to terminal. In addition, microfilaments run through the cytoplasm. Some of these are actin (p. 116), but nerve cells also include specific neurofilaments with a different composition.

Both groups of filaments seem to contribute to intra-cellular transport, and the mainte-nance of shape.

The organization of nervous tissue

We have examined the electrical events taking place at the neuronal cell membrane, and the structure of the cell body and axon. Next, we must examine how these nerve cells are arranged in the body. How is nervous tissue organized?

Nerve cells are in contact with other cells over their entire surface. The areas of membrane not associated with synapses or neuromuscular junctions are in contact with specialized supporting cells. We have seen how axons of peripheral nerves are surrounded by Schwann cells. A similarly close sheath of cells surrounds central axons and cell bodies, except at synapses. The extracellular space around every nerve cell is limited to a narrow cleft between neurone and supporting cell.

This complex of neurones and supporting cells makes up nervous tissue, which is separated from every other tissue, and from the extracellular spaces of the body, by a complete basal lamina. Blood-vessls run in the brain and spinal cord, but they do so in narrow clefts and channels of connective tissue, separated from neurones in nearly every site by a basal lamina and a complete wall of supporting cells. Some molecules and ions diffuse rapidly into nervous tissue, in spite of this separation; for other molecules, including many drugs, there is a significant blood–brain barrier.

The supporting cells of the central nervous system

These make up about half the volume of the central nervous system. They fall into three main classes, named after their appearance in certain silver impregnation techniques, which display the entire cell. They are oligodendrocytes, astrocytes and microglia.

Oligodendrocytes and Schwann cells. These cell types are similar in structure and function, oligodendrocytes ensheathing the axons that lie within the central nervous system, while Schwann cells, as we have seen, surround axons, both myelinated and non-myelinated, in the peripheral nerves.

What, then, is the principal function of the oligodendrocyte? (Note 12.H)

Astrocytes and satellite cells. Astrocytes are the equivalent in the central nervous system of the satellite cells which surround neuronal cell bodies in peripheral collections of neurones, or ganglia. Astrocytes have many branching processes, one or more frequently ending near a capillary (foot processes) and the others spreading out between neurones. The entire surface of each neurone is in contact with other cells in between synapses: astrocyte processes occupy the available space on cell bodies, whereas oligodendrocytes surround the axons. It has been shown that astrocytes are linked by gap junctions. Their membrane is not excitable.

Their precise functions are still debated. One suggestion is that repeated firing of a neurone results in potassium ions accumulating in the 20 nm gap between neuronal membrane and astrocyte: if the level of extracellular potassium were allowed to build up it would eventually interfere with membrane activity in the neurone. Astrocytes are capable of pumping potassium into their cytoplasm without generating an action potential, and the gap junctions allow them to spread the potassium load from areas of high neuronal activity.

Microglia. These, as their name implies, are small cells, which are scattered throughout the central nervous system but have no peripheral equivalent. They appear to be phagocytic.

The epithelial nature of nervous tissue

It is interesting to note that many of the structural features of nervous tissue are similar to those seen in epithelia. Embryologically, the central nervous system develops from the ectoderm. The adult brain and spinal cord contain only a small extracellular interstitial space, being composed almost entirely of cells. These cells are separated by a continuous basal lamina from surrounding tissues. There are neither blood-vessels nor lymphatics in the central nervous system: blood-vessels run in small channels of connective tissue, separated from nervous tissue by the basal lamina, in a way somewhat reminiscent of the fine layers of connective tissue in a compound gland such as the pancreas. There is no collagen in the central nervous system, apart from that surrounding the blood-vessels. Finally, cells of one class, the astrocytes, are linked by gap junctions.

Obviously, there is no external surface like that on to which epithelia face, and no clear polarity of the cells with respect to it. Neurones do not communicate with the rest of the neuro-epithelium by gap junctions, and their axons are capable of growing out of it into other tissues, though even there they remain separated by a basal lamina.

On balance, it is useful to recognize the epithelial characteristics of nervous tissue. The brain and spinal cord are soft and easily deformed, lacking the firmness that comes from a connective tissue skeleton: anatomically, they are protected by the bones of the cranium and the vertebral column, and the fluid that surrounds them and fills the ventricles of the brain.

The histological recognition of central nervous tissue (Fig. 12.12) begins with the realization that one is looking at a massive epithelium. The scattered, large cell bodies of

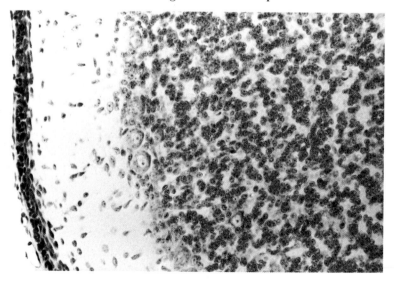

Fig. 12.12   Outer layers of the cerebral cortex, with blood-vessels in the surface (left). Cresyl fast violet. (× 150.)

(a)   (b)

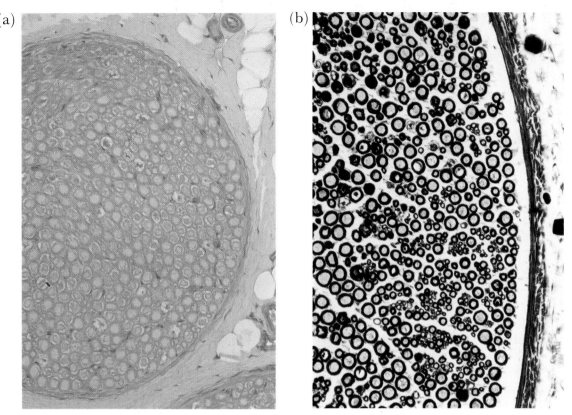

Fig. 12.13   Transverse sections of peripheral nerve. (a) H & E. (× 200.) (b) Osmium fixation. (× 230.)

neurones help to confirm the diagnosis. Remember that nerve cells also occur outside the central nervous system, in ganglia of the autonomic system and on sensory nerves, and in clusters within the walls of the gastro-intestinal tract (Fig. 11.9).

Peripheral nerves

These may be encountered in most tissues and organs. Apart from the major motor and sensory nerves to skeletal muscle, sensory nerves come from many structures in the body, and fine bundles of motor and sensory nerves follow and supply the walls of arteries, and most viscera.

Once an axon leaves the central nervous system, it comes out from the shelter of skull and vertebral column into an environment where movement is the rule. The long, narrow cylinder of cytoplasm is very vulnerable, having almost no resistance to stretching, while pressure can also stop the flow of cytoplasm down the axon and interfere with nerve transmission. Each axon has its own fine sheath of collagen, outside its Schwann cells and basal lamina: this is the endoneurium. Axons are bound together in bundles by thicker bands of collagen, forming the perineurium, while the outer margin of the nerve is made up of an even thicker collagenous layer, the epineurium. Between the nerve bundles in larger nerves run blood-vessels, lying in loose connective tissue whose extracellular fluid and GAGs, contained within the epineurium, provide some protection against local pressure on the nerve. The final result of these protective layers is that nerve axons form a very small fraction of a peripheral nerve (Fig. 12.13). One is looking mainly at an elaborately organized structure of connective tissue containing many Schwann cells, some of which produce myelin sheaths.

With standard histological techniques, what would you expect myelin to look like? (Note 12.1)

The axons themselves are not visible by light microscopy in non-myelinated fibres: in the larger, myelinated fibres, a tiny tag of cytoplasm in the centre of the myelin sheath can usually be seen. Note the very different appearance of a peripheral nerve after fixation in osmium tetroxide, which stains the myelin black and preserves it through histological processing (Fig. 12.13).

Review of communication by means of nerves

In summary, nerve cells release their signalling molecules into the extracellular fluid to diffuse to recipient cells. They convert this slow and imprecise method of communication into a rapid and precise one by extending long processes to within a few nanometres of their target cells and liberating their transmitters into small pockets of fluid of limited volume. The evolution of a mechanism for converting the potential difference across the cell membrane into an action potential has allowed neurones to control transmitter release at distant terminals with extraordinary speed. The integrating function of a single neurone, assessing the total input to it from synapses on its membrane and converting this into its own action potential if it reaches a given level, is the basis for all nervous activity and has been built into circuits of neurones of immense complexity.

Further reading

Keynes, R.D. (1979). Ion channels in the nerve cell membrane. *Scient. Am.* **240:3**, 98–107. Looks at movements of sodium and potassium during the action potential.

Lester, H.A. (1977). The response to acetylcholine. *Scient. Am.* **236:2**, 106–118. Reviews the effects of a neurotransmitter on a recipient cell, in this case muscle.

Morrell, P. and Newton, W.T. (1980). Myelin. *Scient. Am.* **242:5**, 74–89. Its structure, formation and function.

O'Malley, B.W. and Schrader, W.T. (1976). The receptors of steroid hormones. *Scient. Am.* **234:2**, 32–43. Reviews events in target cells for steroid hormones.

Pitts, J.D. (1980). The role of junctional communication in animal tissues. *In Vitro* **16**, 1049–1056. Gap junctions reviewed.

Schwartz, J.H. (1980). The transport of substances in nerve cells. *Scient. Am.* **242:4**, 122–135. Axonal transport reviewed, both from cell body to terminal and in reverse direction.

Stevens, C.F. (1979). The neuron. *Scient. Am.* **241:3**, 48–59. One article in a number devoted to the brain.

Tsukahara, N. (1981). Synaptic plasticity in the mammalian central nervous system. *A. Rev. Neurosci.* **4**, 351–379. An examination of recent evidence for plasticity and its functional importance.

Many textbooks on neurobiology are available. I have found "From Neuron to Brain", S.W. Kuffler and J.G. Nicholls (1976), Sinauer Associates, Sunderland, Mass. to be clear and up-to-date.

References

Cunha G.R. (1976). Epithelial-stromal interactions in development of the urogenital tract. *Int. Rev. Cytol.* **47**, 137–194.

Lehtonen E. (1976). Transmission of signals in embryonic induction. *Med. Biol.* **54**, 108–158.

Wischik, C.M. and Rogers, A.W. (1982). Cell organization in the stroma of the rat uterus. I. The ovariectomised rat. *J. Anat.*, in press.

13
The life and death of cells

The overwhelming majority of histological preparations show us dead cells, killed by fixation. Trying to visualize movement and activity in these complicated patterns is very difficult, like attempting to reconstruct a game of football from one press photograph. Looking at living cells by phase contrast microscopy allows us to see the types of movement that occur rapidly in separated cells or those in culture, and to carry the memory of them over into our observations of fixed tissue. But many transitions and cell migrations occur in living tissue with a much slower time base, and these cannot readily be demonstrated; yet they are essential to the organization and functioning of the tissues of the body. Cell division, cell death and cell migration come into this category.

We all know at an intellectual level that cells divide: each of us begins as a fertilized ovum, a single cell which gives rise to all the complex multitudes of cells that form the adult body. It is equally clear that cells die in the adult body, and are replaced by new cells produced by continuing cell division. It remains difficult nevertheless to appreciate the patterns of cell replacement in the adult body from looking at histological sections of organs and tissues.

Mitosis

The process of division, or mitosis, in which the chromosomes that carry the genetic material are duplicated, distributes one complete set of them to each daughter cell. Mitosis is described in a number of stages (Fig. 13.1). In the first, prophase, the disperse chromatin of the interphase, or non-dividing, nucleus starts to condense into long, fine threads. This condensation withdraws from use the genetic material, which becomes unavailable for transcription of its DNA. Towards the end of prophase, the nuclear membrane breaks down and disappears.

What is the principal function of the nuclear membrane? (Note 13.A)

Though described in stages, mitosis is a continuous process. The condensation and shortening of the chromosomes continues, until, at metaphase, they are short, stubby bodies, staining deeply with haematoxylin and lying approximately in the centre of the cell. By this time the two centromeres have moved to opposite ends of the cell, defining the poles of a sphere on the equator of which lie the chromosomes. A structure called a spindle develops, consisting of fibres which radiate out from each centromere to be attached to each chromosome. At this stage, each chromosome splits in two longitudinally. In the

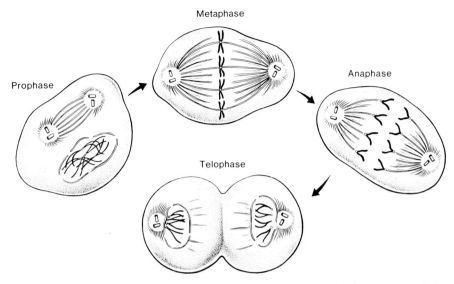

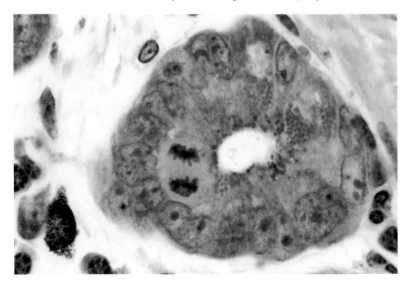

Fig. 13.1   The stages of mitosis, showing the centromeres, the spindle fibres, the chromosomes and, in prophase and telophase, the nuclear membrane.

succeeding anaphase, the spindle fibres shorten, pulling the daughter chromosomes apart and dragging one daughter of each pair towards its corresponding centromere. Finally, in telophase, the chromosomes clustered near each centromere start to become elongated and disperse, and a new nuclear membrane forms around them. At the same time, a cleavage furrow appears in the cytoplasm between the two nuclei: this gradually deepens until it finally separates the cells from each other.

By what mechanism does the cleavage furrow deepen? (Note 13.B)

In normal histological preparations it is usually easy to recognize the stages of metaphase and anaphase (Fig. 13.2), in which the chromosomes are highly condensed and deeply stained, and there is no nuclear membrane. Prophase and telophase are more difficult, unless the material is of high quality. Note that the equator of the spindle usually lies in the centre of the cell. In many tissues, particularly epithelia containing columnar cells, the

Fig. 13.2 Transverse section of crypt from the small intestine of a monkey, showing a cell in anaphase. H & E. (× 1080.)

nucleus is not normally central but basal in position. In these tissues, the presence of metaphase chromosomes centrally — out of line with the other nuclei — makes them very easy to recognize. But while mitotic figures can be seen in a few tissues, in many others they are very rare, even though we know that cell division goes on.

The cell cycle in intestinal epithelium

The technique that has made the most important contribution to our understanding of cell replacement in the body is autoradiography.
What is autoradiography? How does it differ from histochemistry? (Note 13.C)

Labelling methods

In an experimental design known as the tracer experiment, cells are labelled, and their subsequent movements and life expectancy plotted by studying the distribution and amount of label in the tissue at later times. To label a cell adequately for a period of days or even months, it is clear that the label must be attached to some component of the cell which is stable, and does not get passed from one cell to another. It is also obvious that the experiment becomes impossible if all the cells in the body are labelled: the label must be firmly attached to one small group of cells only, if their subsequent fate is to be followed. These two requirements are met by injecting animals with radioactive thymidine (usually labelled with tritium), a precursor of DNA. DNA is to all intents and purposes stable, undergoing no appreciable replacement during a cell's lifetime. It does have a very slow turnover rate, as damage to DNA molecules is repaired, but neurones labelled in the developing brain of a foetal rat are still labelled two years later in old age, and oocytes labelled in the ovaries of foetal mice carry that label into fertilized ova during their reproductive life — effectively carrying the label into the next generation. The second requirement of a good label is also met. An injection of tritiated thymidine into an animal results in incorporation of radioactivity into all those cells which are synthesizing new DNA in preparation for mitosis. Cells which are not synthesizing new DNA do not incorporate it, but break it down to compounds which are excreted: an injected dose is only effectively available for 15–20 minutes. So we can label a fraction of the cells in any tissue — those that are about to divide — and the label is stable.

Phases of the cell cycle

If one prepares an autoradiograph of rapidly dividing tissue, such as the epithelium of the small intestine (Fig. 13.3), taken from a mouse killed a few minutes after the injection of tritiated thymidine, two things are immediately obvious. The metaphase/anaphase figures are not labelled, and are heavily outnumbered by the cells that are labelled. If one examines the same tissue from a whole series of animals killed at increasing times after the injection of ^{3}H-thymidine, metaphases begin to become labelled 1·0–1·25 hours after injection; all of them appear labelled between 1·5 and 8·5 hours (Fig. 13.4). After this, the proportion of labelled metaphase/anaphase figures drops back to zero. A second wave of labelled figures appears 11–13 hours after the first, though they are noticeably less heavily labelled.

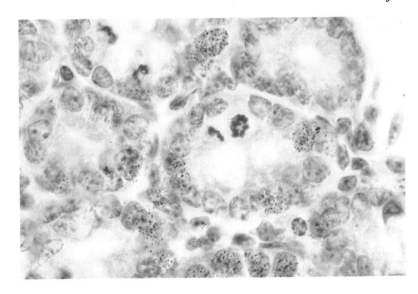

Fig. 13.3 Autoradiograph of crypts from small intestine of a mouse, killed 1 hr after an injection of ³H-thymidine. Haematoxylin. (× 990).

From these data we can work out a great deal about the cell population, given one additional fact. We know from the observation of many cell types in tissue culture that metaphase/anaphase lasts approximately 45 minutes.

Cells in this tissue divide on average once every 11–13 hours, since this is the length of time betwen the first appearance of labelled metaphases and the second: the second wave represents the division of labelled daughter cells from the first wave of mitosis.

The cells that are labelled are those synthesizing DNA while the ³H-thymidine was available. Since it took 1·25 hours for metaphases to become labelled, there is a gap of 1·25 hours between the end of DNA synthesis and metaphase — this is called the G2 period. The labelled cells were presumably spread out randomly through the entire period of DNA synthesis — called the S phase. Some will only just have started the synthesis of new DNA; others will be completing the process. Since it took about 7 hours for this procession of cells to pass the recognizable stage of metaphase/anaphase, it is fair to assume that the S phase lasts about 7 hours.

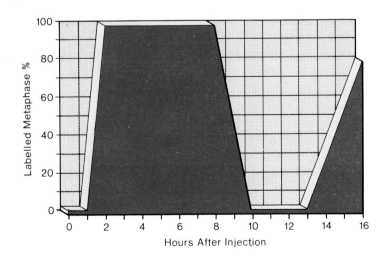

Fig. 13.4 Percentage of labelled mitotic figures in small intestine of mice, injected with ³H-thymidine at various times before killing.

Now, by subtraction, we can calculate the length of time each cell takes to get from the end of anaphase to the beginning of the next S phase — this time is the G1 phase. This will be (the cell cycle time) — (S + G2 + M) = about 3 hours, if we take a figure of 12 hours for the cell cycle.

Relatively simple observations on a series of autoradiographs can give us precise times for a process that could only be partially followed in routine histological material. There are many refinements to this particular experiment and many cells and tissues have been studied. In general, the figures given here for the S, G2 and M phases are much the same whatever the technique and whatever the tissue. The difference between a group of cells that divides rapidly and one that does so slowly is to be found in the length of the G1 phase — in other words, in the rate at which cells enter the S phase. It seems as if cells take a decision to divide on passing from G1 to S: from that moment they are committed to a closely regulated programme that carries them through to the end of telophase in 9·0–9·5 hours.

Cells entering prophase have already doubled their content of DNA. The longitudinal splitting of each chromosome in anaphase represents the redistribution of genetic material, not the stage of its duplication.

We can see now why there are so few metaphases, even in a rapidly dividing tissue like intestinal epithelium. If all the cells in a region of epithelium divide once every 12 hours and metaphase/anaphase lasts 45 minutes, one can expect about 6% of the cells to be in metaphase/anaphase at any one time. By contrast, since the S phase lasts 7 hours out of a total cell cycle of 12 hours, 55–60% of the nuclei will be labelled at short times after injection.

The fate of labelled cells

In the adult, the total number of cells remains approximately constant, so cell division must be balanced by cell death. This balance is often maintained by migration of newly produced cells from the site of mitosis to the site of death. The label introduced into cells at DNA synthesis is ideal for tracing such migrations and calculating life expectancies.

Figure 13.5 shows two autoradiographs of small intestine, one from a mouse killed 1 hour after ^{3}H-thymidine, the other from a similar mouse 36 hours after ^{3}H-thymidine. At 1 hour the only epithelial cells that are labelled are in the crypts, the simple glands that extend down into the wall of the intestine from its surface. The epithelial cells at the necks of the glands and over the surfaces of the finger-like villi which project into the lumen are unlabelled. The only labelled cells in the villi are in the core of connective tissue. By 36 hours, however, most of the cells lining the sides of the villi are heavily labelled. The cells in the crypts are also labelled, but much less intensely.
Why are the crypt cells less heavily labelled? (Note 13.D)
Timed series of autoradiographs show that the epithelium lining the villi is replaced every 36–48 hours in the mouse. New cells migrate up from the crypts, pushing the old ones before them towards the tips of the villi, where they fall off into the lumen of the intestine. Here they are digested, and their constituents are reabsorbed into the body through the cells that have replaced them.

Dividing cells are very vulnerable, and sites of cell division in the body are often protected from the harsher environments that produce cell death. It follows that migration

(a)   (b)

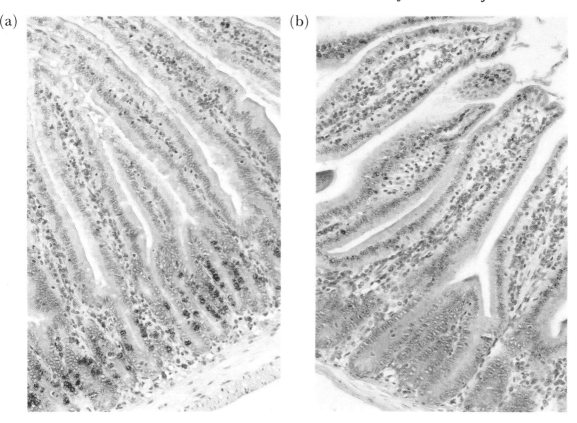

Fig. 13.5   Autoradiographs of small intestine from two mice injected with ³H-thymidine. (a) Killed 1 h after injection. (b) Killed 36 h after injection. Haematoxylin. (× 200.)

of newly produced cells away from their place of mitosis towards their functional site occurs in many tissues. The same experimental approach that we have applied to intestinal epithelium can be used in any tissue to trace the fate and life expectancy of cells, providing it is possible to introduce radioactive thymidine at the appropriate time at which DNA synthesis occurs. Such labelling experiments have resolved many long-standing arguments between histologists, showing for instance, the possible fates of the monocyte circulating in the blood stream, and the complex circulation of lymphocytes.

What does become of the monocyte? What, briefly, is the pattern of circulation of lymphocytes? (Note 13.E)

Patterns of cell renewal

Many cells and tissues have now been studied from the viewpoint of cell renewal, and four major patterns have emerged. Some cell types, like the crypt cells of intestinal epithelium, are constantly dividing; the rate at which they divide varies widely. The second group does not normally divide but can do so to make good losses from cell death. The third group does not divide, even to replace lost cells. The fourth, rather specialized group, comprises the male and female germ cells, which undergo meiosis, the reduction division which produces gametes with half the normal complement of DNA and of chromosomes: fusion of two gametes at fertilization gives a new individual with the full content of DNA and chromosomes per cell.

Continually dividing populations

This pattern is seen in many epithelia which, facing an uncontrolled external environment, have a continuing rate of cell loss. Apart from the gastro-intestinal tract, replacement rates are not remarkably high, and it requires a bit of searching to find a metaphase figure in the basal layer of the skin. Epithelial areas with fairly high replacement rates occur, for instance, in the roots of hairs; those with very low replacement rates face relatively mild environments, such as the duct system of the pancreas.

Apart from epithelia, blood cells have a finite life and need regular replacement. The polymorphonuclear leucocytes are the ones with the shortest survival, with a mean life of 3–5 days after release from the bone marrow. Red blood cells, by contrast, survive 100–120 days. The bone marrow is the site of continuous, active cell division throughout life.

As always with complex biological systems, classification is not easy, particularly in some of the more slowly dividing tissues. The distinction between such tissues and those that only divide in response to cell loss is not clear cut.

Cells that divide in response to cell loss

The much-quoted example in this group is the liver. The epithelial cells of the liver are highly specialized and do not normally divide, a state known as the G0 phase. If, however, one half of the liver is surgically removed from an experimental animal, a wave of cell division occurs which ultimately restores the liver to its original mass. Once this is achieved, cell division ceases again. Liver damage from viral infections or alcohol is also associated with nodules of newly produced cells.

In a sense, all cell division is geared to cell loss in the adult, and it is difficult to define the difference between these cells and ones that divide continually. It is arguable that they are similar, except for the level of cell loss to which they normally respond.

Cells that never divide

There are cell populations which clearly differ from those discussed above, since they do not divide in the adult, even in the presence of massive cell loss. The prime example of this pattern of behaviour is the neurone. In the developing central nervous system, cell division ceases in neurones at or soon after birth, though it continues in glial cells.
What are glial cells? What types are there and what are their principal functions? (Note 13.F)
It has been calculated, I do not know with what precision, that an adult human loses neurones from the central nervous system at a rate of about 10 000 per day. The original population is so vast and there is such inbuilt plasticity in neuronal function that we seem able to tolerate this loss without great effects on function. If the loss is significantly speeded up by injuries repeated over several years, as in an unsuccessful boxer, functional loss becomes apparent. This continued loss of cells, even when accompanied by loss of function, never produces a compensatory mitosis in the population of neurones.

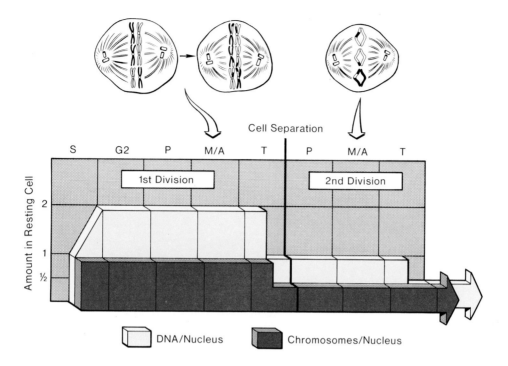

Fig. 13.6   The recognizable stages of meiosis (above) correlated with the number of chromosomes per cell and the content of DNA per cell (below).

The regulation of cell division

The cell populations that divide continually and those that respond to injury by division produce cells at a rate very closely matched to the body's requirements. This is particularly well illustrated following wounding to an epithelium such as the skin, or the removal of part of the liver, quoted above. Cell division increases locally in the margins of a skin wound, returning to basal levels as soon as the defect is repaired, in the same fashion that the liver ceases to grow when it reaches its initial mass. Similar effects can be seen in the bone marrow following a haemorrhage, with the production of more red cells than usual until the loss is repaired. The nature of the processes that regulate cell division has been the subject of intense speculation and much experiment. One proposed mechanism with many adherents states that mature, differentiated cells produce inhibitors of cell division, specific to cells of their own type, and that loss of mature cells results in a local reduction in the levels of those inhibitors (which are called chalones), thus permitting cell division to occur (Bullough, 1969). Such chalones may contribute to the regulation of cell division in some circumstances, but they certainly are not the whole story. In many instances, we have clear indications that cell division is linked to the functional capacity of the tissue, rather than to the number of mature cells. Moving to live at high altitude results in oxygen tensions in the tissues much lower than at sea level: such a move is a potent stimulus for the production of more red blood cells, until oxygen tensions in the tissues rise to normal again. This increased mitotic rate is associated with higher circulating levels of erythropoietin, which stimulates cell division — not with reduced levels of an inhibitory chalone.

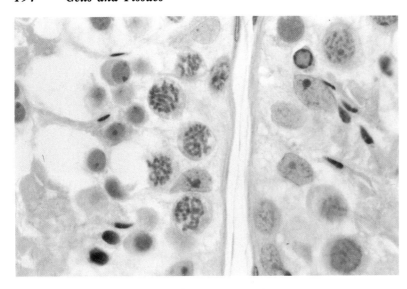

Fig. 13.7   The lining of two adjacent tubules in the testis: note the large number of dividing cells. H & E. (× 960.)

The male and female germ cells

The germ cells are populations produced by mitotic divisions of precursor cells, which, in their final differentiation, pass through the specialized process of meiosis. This results in gametes which have half the normal complement of DNA and half the normal number of chromosomes. Though this pattern of population growth by mitosis, followed by meiosis and differentiation, is common to both male and female germ cells, the details and timing of the process are quite different in the two sexes.

Meiosis

The process of meiosis spans two successive cell divisions, and is illustrated diagrammatically in Fig. 13.6. The synthesis of new DNA precedes the prophase of the first division, in much the same way that the S phase comes before mitosis. Each chromosome that appears during this first prophase is thus effectively double, though it appears down the microscope as a single body. During the first metaphase, these chromosomes arrange themselves at the equator of the cell and a spindle forms: unlike the sequence in mitotic metaphase, however, the chromosomes arrange themselves at the equator in pairs, homologous chromosomes, from the maternal and paternal sets inherited at conception, lying side by side. Each chromosome is already two chromatids, so chromosome pairs are often called tetrads. While lying together, chromosomes may exchange genetic material with the other member of the pair: crossing over, as this is called, takes place between the inner pair of chromatids in the tetrad (Fig. 13.6). As the chromosomes move apart in anaphase of the first meiotic division, each daughter cell acquires a mixture of chromosomes, some deriving from the maternal set and some from the paternal. This mixing of chromosomes and the crossing over process are mechanisms for increasing genetic diversity in the offspring. The first telophase sees the reconstitution of two daughter nuclei. These nuclei have half the number of chromosomes found in an ordinary somatic cell but, since each chromosome is already double, they have the normal, characteristic amount of DNA.

The second meiotic division is not preceded by DNA synthesis. At the second metaphase, the reduced number of chromosomes comes to lie at the equator of the spindle,

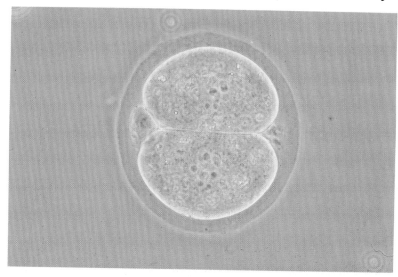

Fig. 13.8 A mouse embryo at the two-cell stage: note the two, small polar bodies lying with the large cells inside the capsule, or zona pellucida. Interference contrast microscopy. (× 290.)

and each splits in half longitudinally, one half moving towards each pole of the spindle. The daughter cells of this division have half the resting number of chromosomes and half the normal content of DNA.

The first meiotic division is often called the reduction division because the number of chromosomes per cell is halved, while the second is referred to as a "mitotic division". This is confusing and inaccurate and should be avoided. Both are reduction divisions, the first halving the number of chromosomes, the second again halving the DNA content per cell. Meiosis is a complex sequence involving two divisions, each unlike a normal mitosis.

Male germ cells

While in mammals the male and female germ cells both undergo meiosis, they differ in just about every other respect. In the male, the epithelium lining the tubules of the testis is the site of continued cell division (Fig. 13.7) from puberty right through to old age: the precursor or stem cells — the spermatogonia — have a cell cycle time comparable to that of the crypt cells of intestine, dividing by mitosis to give a very large and rapidly renewing population of spermatocytes. These enter meiosis, each spermatocyte giving rise to four sperm as a result of the two meiotic divisions. So male gametes are produced in very large numbers continually, and are relatively short-lived. Labelling experiments show that about 30 days elapse between the final stage of DNA synthesis and the appearance of labelled sperm in the ejaculate.

What will be the latest stage in the production of sperm at which ³H-thymidine can be incorporated into them? (Note 13.G)

Female germ cells

In the female, the stem cells which are the equivalent of spermatogonia are called oogonia. They divide mitotically in the ovaries to produce a population of oocytes, cells capable of entering meiosis. In extreme contrast to the male, oogonia cease to divide before birth, so that the ovaries of the late foetus contain no oogonia, but only oocytes. The latter synthesize DNA and enter the prophase of the first meiotic division before birth, but here

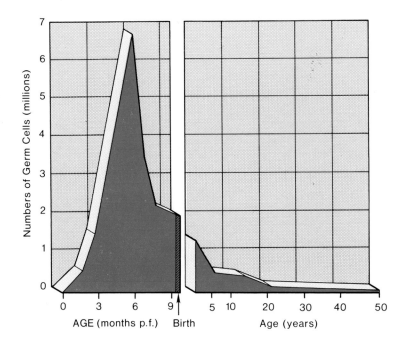

Fig. 13.9   A graph of germ cell numbers in the human female through pre-natal and post-natal life.

they stick, in a sort of suspended animation. From before birth to puberty, the ovaries contain oocytes in this resting stage of the first meiotic prophase. During this time the population of oocytes is dramatically reduced by cell death: in the human female, it falls from a peak of 7×10^6 before birth to about 150 000 at puberty (Baker, 1963). At ovulation, an oocyte is shed from the ovary to enter the uterine tubes: ovulation coincides with the resumption of the first meiotic division. The spindle for this first division is not centrally situated in the cell, but adjacent to the cell membrane, and the process produces an unequal division of the cytoplasm, most of it remaining with one daughter cell which is still called the oocyte, the rest forming a small cell, the first polar body. The stimulus for the second meiotic division is penetration of the oocyte membrane by a sperm at fertilization: this, too, is an unequal division, resulting in a second polar body (Fig. 13.8). Meiosis in the male produces four gametes, in the female only one.

Labelling of the population of oocytes, therefore, has to take place in the foetal ovaries before birth, and, in the human, the time that elapses between this and the appearance of a labelled gamete in the reproductive tract is 13–45 years. Oocytes, like neurones, form a population that cannot renew itself. The death rate is illustrated in Fig. 13.9, and it results in the total elimination of the population by the end of reproductive life, the menopause.

Summary

In conclusion, the static picture we get of cells and tissues from histological sections is incomplete and misleading. Cells form living populations, with life expectancies varying from a few days to the entire lifespan of the animal. Renewal of some populations occurs, and is very accurately regulated to the functional needs of the body. Other populations do not replace themselves, but decline progressively with time. It is clear that injury to such a population will not be repaired, while renewing populations, by contrast, can adjust their rate of division to make good unexpected losses.

It is unfortunate that standard histological sections do not remind us more clearly about this beautifully regulated renewal of cell populations.

Further reading

Mazia, D. (1974). The cell cycle. *Scient. Am.* **230:1**, 54–64. The place to start: clear and well illustrated.

Potten, C.S. (1981). Cell replacement in epidermis (keratopoiesis) via discrete units of proliferation. *Int. Rev. Cytol.* **69**, 271–318. A detailed and up-to-date look at epidermal renewal.

Yanishevsky, R.M. *et al* (1981). Regulation of the cell cycle in eukaryotic cells. *Int. Rev. Cytol.* **69**, 223–259. A recent review of the control of cell division.

References

Baker, T.G. (1963). A quantitative and cytological study of germ cells in human ovaries. *Proc. R. Soc. B.* **158**, 417-433.

Bullough, W.S. (1969). The control of tissue growth. *In* "The Biological Basis of Medicine", Vol. 1. Bittar, E.E. and Bittar, N. (Eds). Academic Press, London and New York.

The small intestine

We have now studied all the major components of living tissues and organs. In this chapter, I want to show how this information can help in the interpretation of histological sections, using the small intestine as an example. It is possible for you to take sections of small intestine, or of any other mammalian organ for that matter, and to make quite detailed and sophisticated deductions about how the organ functions. The reverse is also true — given basic information about the physiology and biochemistry of an organ, its histology can largely be predicted. In practice, you will seldom have to do either task quite as I have described them: you will, however, find histology and physiology, structure and function, reinforcing each other, so that both become more logical and easier to learn and to remember.

Initial observations on sections of small intestine

A transverse section of the jejunum is clearly from a hollow tube, with a reasonably large diameter (Fig. 14.1). A glance at slightly higher magnification (Fig. 14.2) shows two fairly thick layers of smooth muscle at right angles to each other.
What can we deduce from the large diameter of the lumen, and the arrangement of muscle in the wall? (Note 14.A)
We would expect a compound rather than a simple epithelium. However, the epithelium is simple columnar (Fig. 14.3).
What can we deduce from the presence of a simple epithelium? (Note 14.B)
There is also a thin, continuous layer of smooth muscle just below the epithelium, the muscularis mucosae (Fig. 14.6), the layer that is characteristic of the gastro-intestinal tract.

The small intestine propels its own contents, and also is the site of active exchange between tissues and lumen. As we have seen in Chapter 11, exchange is most efficient in thin-walled tubes of small diameter. But the intestine needs to propel down its lumen relatively large meals, which requires a large lumen and a muscle coat of two or more layers. In such a tube, efficient absorption from the lumen is only possible if a number of specializations are present to counteract the effects of the large lumen and the thick wall of the tube. These specializations have three main aims:
1. To increase the surface area exposed to the luminal contents.
2. To keep the luminal contents continually mixed and stirred.

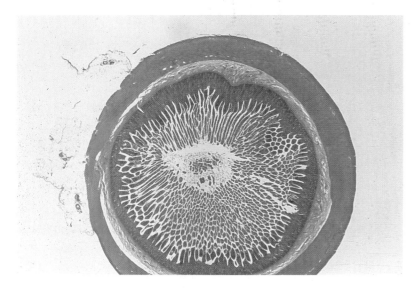

Fig. 14.1   A transverse section through the jejunum. H & E. (× 10.)

3. To bring blood and lymphatic capillaries very near to the epithelium, so that absorption does not have to take place through the full thickness of the intestinal wall.
   We will look at the jejunum in sequence from lumen outwards bearing these in mind.

The region inside the muscle layer

This comprises mucosa and submucosa, separated by the muscularis mucosae (Fig. 11.7).

The mucosa

At low power (Fig. 14.2) the mucosa is thrown into a number of folds, which can be seen naked-eye as longitudinal ridges on the luminal surface of the fresh intestine. In addition,

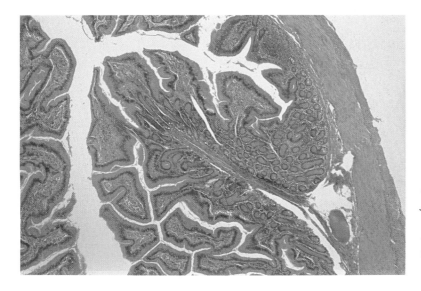

Fig. 14.2   The wall of the jejunum: serosa and muscle (right). Note the large mucosal ridge (centre) and the many, smaller villi. H & E. (× 40.)

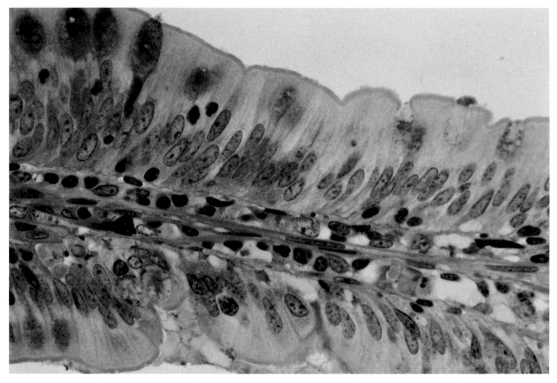

Fig. 14.3   Longitudinal section through a jejunal villus. H & E. (× 640.)

the whole inner surface of the small intestine carries finger-like projections, called villi, an order of magnitude smaller than the mucosal folds.

The epithelium lining the lumen covers the villi, and is also continuous into numbers of simple glands which extend from the surface of the small intestine down into its wall (Fig. 13.5). It may not be altogether easy to define the level which represents the notional surface from which villi project and on to which glands open.

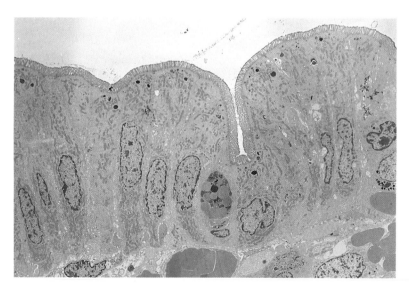

Fig. 14.4   TEM of villous epithelium. (× 1500.) (Courtesy, Dr M. Haynes.)

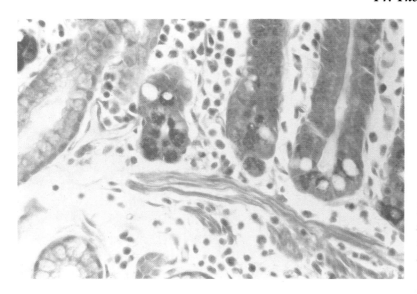

Fig. 14.5 Paneth cells with magenta-stained secretion granules, in the bases of duodenal crypts of monkey. Masson's trichrome. (× 405.)

Over the villi, the columnar epithelial cells are of two distinct types (Fig. 14.3). The more common cell, the enterocyte, has a fairly large nucleus, lying towards the base of the cell, oval in shape and with relatively little condensed chromatin. Its cytoplasm is not noticeably basophilic and there are no obvious secretion granules. The surface facing the lumen has the fuzzy appearance called a brush border. There are no obvious striations at the base of the cell, at right angles to the basal lamina.

What deductions can you make about the principal functions of these cells from this description? (Note 14.C)

Seen with the TEM (Figs 14.4, 4.11), enterocytes have a luminal surface covered with microvilli closely packed together, below which they are joined to their neighbours by a junctional complex. Below this again, the lateral cell membranes of adjacent cells are not closely bound together, but face a wedge-shaped extracellular space lying on the luminal side of the basal lamina. They have only a little RER and the Golgi apparatus is not large. There are a considerable number of mitochondria and many small vesicles, suggesting transport, but nothing to imply phagocytosis on a considerable scale.

The second cell type in the epithelium is the mucus-secreting goblet cell, well demonstrated by histochemical techniques for glycoproteins (Fig. 4.10), but quite recognizable in sections stained with H & E from its characteristic shape, with mucus bulging the luminal half of the cell and pushing the nucleus towards the base (Fig. 4.4).

In the epithelium of the glands, or crypts of Lieberkuhn, the two cell types found over the villi can also be seen in the necks or upper quarter of the glands. Below this, the striking feature of the epithelium is the high probability of seeing metaphases and anaphases. Cell replacement in this epithelium has been discussed in Chapter 13.

Apart from the metaphase/anaphase figures, what other structural feature reflecting this high turnover rate might you look for in a routine section stained with H & E? (Note 14.D)

At the base of each crypt, below the region of rapid mitosis, a cluster of rather different cells can be seen (Fig. 14.5). These, called the Paneth cells, have all the characteristics associated with the intermittent secretion of proteins, including cytoplasmic basophilia and secretion granules clustered at the cell apex. Their precise function is unknown: suggestions that have been made include the production of anti-bacterial substances.

Summarizing the epithelium alone, it is specialized for absorption, with mucosal folds and villi greatly increasing the surface in contact with the lumen and the microvilli of the enterocytes doing the same job at a different level of magnification. The surface layer of mucus acting as a protective barrier, and the very rapid rate of cell replacement from specialized crypts, both indicate a harsh environment in the lumen and a rapid rate of cell death.

The lamina propria. This layer extends from the basal lamina under the epithelium to the muscularis mucosae. It varies in structure with its position — that in the villi differs from that around the crypts. In the villi it is very loose connective tissue, with many capillaries lying beneath the basal lamina and many obvious lymphatic vessels (Fig. 14.3).

In the villous core there are also large numbers of lymphocytes, some of which have passed through the basal lamina into the spaces between the basal ends of enterocytes; plasma cells can also be seen in considerable numbers. In places, nodules of lymphoid tissue occur in the lamina propria, complete with reticulo-endothelial cells.

How do you interpret this abundance of lymphoid cells in this site? (Note 14.E)

Running longitudinally in the centre of the villous core there are several smooth muscle fibres. These reach the muscularis mucosae.

The connective tissue around the crypts is less remarkable, resembling that found in many glands: it contains collagen and a basket-like capillary plexus surrounding each gland.

Absorption of food materials from the lumen of the intestine transfers sugars and aminoacids into the extracellular fluid immediately beneath the basal lamina under the epithelium of the villi. This is precisely where the rich capillary network of the villi is found, with fenestrated endothelium to assist the entry of absorbed material into the blood. Fats are absorbed into the enterocytes as fatty acids and reassembled into lipid droplets in their cytoplasm, to be subsequently secreted into the spaces between the basal ends of adjacent cells. From here, they pass through the basal lamina into the villus, but are unable to enter the capillaries. They do, however, find their way into the lymphatic vessels.

By what mechanism can particulate matter pass from the extracellular spaces into lymphatic vessels? (Note 14.F)

The simple epithelium and the very short distance between basal lamina and capillary make possible the efficient transfer of absorbed material into the blood stream in the innermost layer of this tube, well within the muscle coat. The small intestine is a thin-walled tube with simple epithelium permitting the absorption of material from the lumen. This is situated inside a thick-walled, muscular tube capable of propelling the contents down the lumen.

The smooth muscle of the villi seems to have two functions. First, it can cause some movement of the villi, helping to stir and mix the contents of the lumen. Secondly, it raises the pressure within the villous core when it contracts, which may be an important factor in moving lymph down the lymphatic capillaries.

The lymphoid tissue and the frequent single lymphocytes and plasma cells indicate that the lumen is not sterile, and that the entry of micro-organisms and foreign proteins is possible through breaches in this delicate epithelium, which faces such a hostile environment.

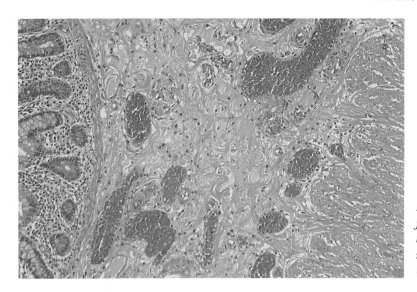

Fig. 14.6 Submucosa of the jejunum. Muscularis externa, to right, mucosa and muscularis mucosae to left. H & E. (× 100.)

The muscularis mucosae and submucosa

The muscularis mucosae is a thin sheet of smooth muscle circling the intestine. Such a sheet can alter the diameter of the lumen when it contracts, but is not capable unaided of propelling the contents down the lumen. As we saw above, the smooth muscle of the core of each villus makes contact with this sheet.

The submucosa, lying between the muscularis mucosae and the muscularis externa, is loose connective tissue which carries arteries, veins and lymphatic vessels to serve the lamina propria (Fig. 14.6). It lacks the elaborate capillary network seen in the villi and around the crypts. It also contains many clusters of large, basophilic cells, each surrounded by smaller satellite cells. The large cells sometimes can be seen to continue into bundles of fine nerve fibres. This is the submucous nerve plexus, or Meissner's plexus: it supplies the epithelium, and may also assist in regulating the activity of the muscularis mucosae and the smooth muscle of the villi.

Most of the tubes in the body that have smooth muscle in their walls are innervated by neurones in the ganglia of the sympathetic or parasympathetic nervous systems; integration of the activity of this muscle with the needs of the other tissues is carried out either in these ganglia or in the central nervous system itself. The gastro-intestinal tract is remarkable in having its own nervous tissue, complete with sensory and motor neurones and inter-neurones, located within its walls. The presence of this enteric nervous system implies greater complexity and independence of function: seeing neurones in the submucosa should immediately suggest to you that the activities requiring this detailed, local control are more varied than simple propulsion or the regulation of flow.

Nerve fibres originating in neurones in autonomic ganglia or in the central nervous system do enter the wall of the intestine, but do not usually innervate smooth muscle, except at sphincters. Instead, they synapse with the intestine's intrinsic neurones, influencing their activity in response to the needs of the body as a whole.

The muscularis mucosae is capable of independent contraction and is not an outlying part of the muscularis externa. Together with the muscles of the villi, it contributes to movements which mix the contents of the lumen and empty the lymphatics of the lamina

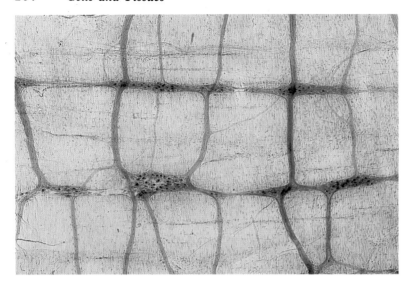

Fig. 14.7 A spread of small intestine, with the myenteric plexus displayed. Silver impregnation. (× 40.)

propria. The very loose connective tissue in the plane bordering on the muscularis externa permits movements of the whole mucosa on the muscle coat.

The muscle layer and the region outside it

Turning now to the muscularis externa and the region outside it, we are looking at the structures responsible for propelling the luminal contents and the adaptations required to accommodate those movements.

The muscularis externa

The muscularis externa of the small intestine consists of smooth muscle arranged in two layers at right angles to one another. Such an arrangement is seen in many tubes of the body which propel their contents. The striking difference between the gastro-intestinal tract and other tubes which propel is the presence of an intrinsic nervous system, complete with neuronal cell bodies and synapses (Figs 14.7, 11.9). This plexus is extraordinarily elaborate: it has been calculated that it contains more neurones than the spinal cord. The neurones appear as small clusters, lying between the inner circular and outer longitudinal layers of muscle.

Like the submucous plexus, this myenteric plexus (Auerbach's plexus) receives extrinsic fibres, but is capable of producing complex sequences of activity in the absence of any nervous input from outside the intestine.

What are the complex movements that require such elaborate local integration? First, there is propulsion of the contents down the lumen. If we look at the ureter, it transfers urine from the kidney to the bladder by peristalsis in a smooth, rapid movement; similarly, the ductus deferens rapidly transfers fluid rich in spermatozoa from testis to urethra. At similar speeds, food would travel the length of the small intestine in minutes, and absorption would be inefficient and incomplete. Clearly, other movements must also occur.

Direct observation of live intestine shows a remarkable variety of muscular activity. At times, rings of contraction will appear in the wall of the intestine, preventing the passage of luminal contents, while waves of less powerful contraction will move in either direction between two stationary rings. Then the rings will relax and new rings appear, and the process will be repeated. At other times, organized waves of contraction will pass for short distances down the intestine, preceded by short zones of relaxation. The latter pattern clearly moves the contents, while the former sequence of movements is able to mix the contents thoroughly without net transfer down the lumen. The muscularis externa, then, has an important function in keeping the luminal contents mixed and stirred, assisted by the activities of the muscularis mucosae and the muscle of the villi. As we have already noted, this constant mixing is essential if efficient transfer of materials from lumen through epithelium to tissue fluids is to take place in a tube with a large luminal diameter.

The serosa

This is a thin layer of connective tissue carrying blood-vessels and nerves to supply the intestine, bounded by a mesothelial layer of cells (Fig. 11.8a). The presence of this serosa rather than an adventitia should not surprise you after the description of the complex and continual movements which characterize the small intestine.

Summary of function and structure of the jejunum

The jejunum is a tube of large diameter with muscular walls, responsible for the propulsion of its own contents. It is also the site of massive absorption of food materials from the lumen into the body. In spite of the large luminal size, absorption is very efficient as a result of several specializations.

(i)   A large surface area is provided by mucosal folds, villi and microvilli.

(ii) Continual mixing of the contents is produced by the activities of muscularis externa, muscularis mucosae and the muscle fibres in the villi. The complicated sequences of contraction needed to achieve this mixing are regulated by the myenteric and submucous plexuses of neurones, and the movements of intestine relative to other structures are accommodated by the peritoneal cavity.

(iii) In the structures where absorption takes place, the villi, the blood and lymphatic capillaries into which absorbed material is taken lie immediately beneath the simple columnar epithelium: the endothelial cells of the blood capillaries are fenestrated to assist in the rapid transfer of materials from extracellular fluid to blood.

The environment in the lumen of the jejunum is very hostile with digestive enzymes acting on the luminal contents, and yet efficient absorption requires a delicate, simple epithelium. Several specializations are present to protect the body against the possibility of destruction of epithelial cells.

(i)   The luminal surfaces of the epithelium are covered with a thin layer of mucus, produced by goblet cells in the epithelium.

(ii) Epithelial cells are continually replaced. Cell division in the protected environment provided by glands (the crypts of Lieberkuhn) balances cell loss from the tips of the villi.

(iii) Many lymphoid modules are present in the mucosa to deal with foreign material, including micro-organisms, which succeeds in penetrating the epithelium. Free lymphocytes and plasma cells are present in great numbers throughout the mucosa.

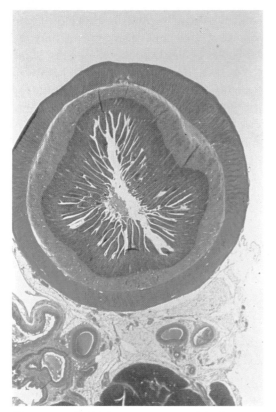
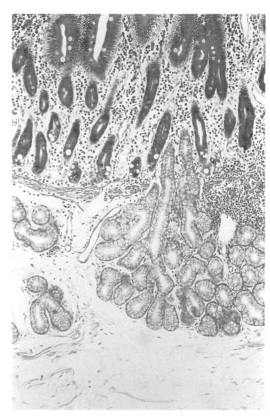

Fig. 14.8   (left) Transverse section of duodenum, with small pieces of pancreas, of blood-vessels and of bile duct below. H & E. (× 10.)

Fig. 14.9   (right) Wall of duodenum of a monkey. Mucosa occupies the upper half, separated by muscularis mucosae from the submucosa (below). Note the submucosal glands. Masson's trichrome. (× 100.)

(iv) The Paneth cells at the bases of the crypts may secrete anti-bacterial substances which would help to maintain the sheltered environment needed for cell division: this, however, is still only a suggestion.

Variations in structure down the length of the small intestine

The other portions of the small intestine, the duodenum and the ileum, have a basic structure similar to that of the jejunum, with villi, crypts, a muscularis mucosae and the two nerve plexuses which control the mixing and propelling movements essential to efficient digestion and absorption. They can nevertheless be distinguished from the jejunum by their histological appearances, which in turn suggests there are some differences in function.

The duodenum

This (Fig. 14.8) receives the food from the stomach intermittently, in fluid form, as it squirts through the pyloric canal. This fluid is very acid and contains the proteolytic enzyme, pepsin. Clearly, the mucosa of the duodenum requires protection against this potentially damaging fluid. The duodenum also recevies the digestive enzymes secreted by the massive exocrine gland, the pancreas, a compound alveolar gland.

What do these terms exocrine, compound and alveolar, mean? (Note 14.G)

These enzymes are secreted by the pancreas as inactive precursor molecules, which are converted to active digestive enzymes in the lumen of the small intestine, enzymes which have an optimum pH well to the alkaline side of that of gastric juice.

The common bile duct, carrying bile salts and pigments from the liver, also opens into the duodenum in company with the main pancreatic duct. This opening is about one-third of the way from the start of the duodenum at the pylorus to its ending at the duodeno-jejunal junction.

If we now try to predict the structural changes that might be needed to accommodate these functional demands, it is clear that the duodenum cannot be permitted the wide freedom of movement within the peritoneal cavity that is possible in the jejunum. The bile duct and pancreatic duct must not be twisted or kinked. The duodenum is the only part of the small intestine to have a fixed position, largely excluded from the peritoneal cavity.

How might one recognize this from a transverse section of duodenum? (Note 14.H)

Next, the acid fluid from the stomach must be neutralized and even made alkaline to protect the duodenal mucosa and to provide a suitable environment for the pancreatic enzymes. This is achieved by the secretions of glands in the walls of the duodenum, which pour out a watery, highly sulphated mucus on to the surface of the epithelium to coincide with the emptying of the stomach. However, the mucosa is already packed with glands, the crypts of Lieberkuhn, which provide the sheltered environment needed for cell division. So the mucous glands lie deeper in the submucosa (Fig. 14.9).

How would you describe them? (Note 14.I)

These are the major structural differences between duodenum and jejunum — the presence of an adventitia rather than a serosa, and the glands (Brunner's glands) in the submucosa. There are other, less obvious, differences. Since the material coming from the stomach is practically sterile, the probability of invasion by micro-organisms in the lumen is low. Lymphoid tissue is less obvious than in the jejunum. Probably as a result of the presence of the submucosal glands, there appears to be less need for the rather thick mucus produced by goblet cells, which are less numerous than in the jejunum. These relative differences require some experience to recognize, unless a section of jejunum is available for immediate comparison, whereas the adventitia and Brunner's glands are easy to identify.

The ileum

This portion of the small intestine is not clearly demarcated from the jejunum. Instead, there is a gradual change in the character of the small intestine from the start of the jejunum to the junction of ileum with large intestine.

In normal circumstances all digestion and most of the absorption of digested materials are complete before the end of the jejunum. The ileum is a functional reserve, available for times of very large but widely spaced meals. As the food is absorbed, so is much of the water that was secreted into the lumen of the gastro-intestinal tract by salivary glands, stomach, pancreas, liver and Brunner's glands. The contents of the lumen are much less bulky in the ileum and consist principally of indigestible materials such as cellulose.

The ileum opens into the large intestine which houses a vast population of micro-

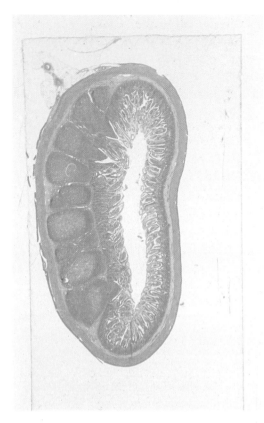
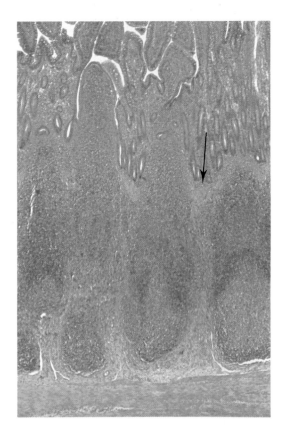

Fig. 14.10   (left) Transverse section of ileum, showing many Peyer's patches to the left of the lumen. H & E. (× 10.)

Fig. 14.11   (right) Peyer's patches in the ileum, extending on both sides of the muscularis mucosae (arrow). H & E. (× 40.)

organisms, and the ileo-caecal valve is not an absolute barrier to their migration. So the bacterial count rises steadily from pylorus to ileo-caecal junction.

Looking at the structure of the ileum (Fig. 14.10), it is very similar to that of jejunum. The villi may be slightly smaller, and the epithelium over the villi contains a higher percentage of goblet cells than in the jejunum — a feature presumably related to the more solid contents and the higher concentration of undigested material. To recognize these features requires some experience, unless a section of jejunum is available for immediate comparison.

The feature that is obvious is the lymphoid tissue. Not only are there more lymphocytes and plasma cells in the lamina propria, but the nodules of lymphoid tissue are larger and more frequent. In fact, many of them reach such a size that they can be seen on naked-eye inspection of the fresh mucosa. Such large accumulations of lymphoid tissue are called Peyer's patches. Histologically (Fig. 14.11), as well as small nodules lying in the lamina propria, there are large masses of lymphoid tissue extending from the epithelium itself right through the muscularis mucosae into the submucosa: these latter are Peyer's patches. Such lymphoid masses affect the structure of the mucosa, which lacks villi or only carries short and abnormally shaped villi over a Peyer's patch.

Summary

No new tissues have been introduced in this chapter. Instead, drawing on the material in the earlier chapters, we have looked at the histology of one organ, the small intestine, seeking out possible correlations between structure and function. Your knowledge of cells and tissues is now good enough to allow you to build similar correlations in any organ you examine. The patterns you seen on microscope slides indicate the ways in which the organ functions in life.

Further reading

Berg, R.D. (1980). Mechanisms confining indigenous bacteria to the gastro-intestinal tract. *Am. J. Clin. Nutrition* **33**, Suppl. 11, 2472–2484. Competition with other bacteria within the lumen is surprisingly important, assisting the immune system to protect the body against invasion.

Gershon, M.D. (1981). The enteric nervous system. *A. Rev. Neurosci.* **4**, 227–272. An up-to-date survey of knowledge on this system.

Granger D.N. (1981). Intestinal microcirculation and transmucosal fluid transport. *Am. J. Physiol.* **240:5**, 343–349.

Moog, F. (1981). The lining of the small intestine. *Scient. Am.* **245:3**, 116–125. An excellent look at the structure and functions of the epithelium.

Interpreting abnormal structure

It is a debatable point where normality ends and the study of disease begins. Each of us at any time has a number of relatively minor scratches and bruises, which are being protected and repaired by the body's "normal" defence mechanisms. We have met these protective mechanisms several times already, notably in Chapter 8. In this chapter, we shall start by examining the responses of the body to injury in the processes of inflammation and repair. Only when we see the polymorphonuclear leucocytes, the macrophages, the plasma cells and lymphocytes co-operating in response to injury can we fully appreciate their particular functions. The second part of the chapter will look at some examples of abnormal structure, partly as exercises in working out the links between structure and function, using unfamiliar material, and partly to extend our understanding of normal structure by observing instances in which the normal control mechanisms go wrong.

Inflammation

Describing inflammation is rather like describing a game of football. The players are in reasonably standardized positions at the start of the game, and the objectives of play remain the same throughout, but each game differs from all other games in the patterns of play that develop. The features basic to the inflammatory response can be modified in various ways. The cause of the damage — a blow, crushing, infection by bacteria or by viruses, burning by heat or sunlight for example — can modify the inflammatory response. The precise pattern of the body's response to injury varies also with the body itself — its age, its nutritional state, its genetic constitution or its affliction with other diseases or abnormalities. And yet, even though the detailed progress of the body's response to injury is almost infinitely varied, certain fundamental patterns can be traced in practically every case. These are collectively called the inflammatory process.

Many factors contributing to the defence of the body circulate in the blood stream and do not normally leave the capillaries in great amounts. Antibodies are a component of the plasma proteins; platelets and many factors involved in clot production are confined to the circulating blood; polymorphonuclear leucocytes, monocytes and lymphocytes do not usually leave the blood vessels for the tissue spaces in large numbers, if we exclude lymphocytes entering lymphoid tissue. The early stages of inflammation involve the mobilization of these and other components into the damaged area. The second stage of responses to injury attempts to restrict damage to the affected area, and to destroy or

(a) (b)

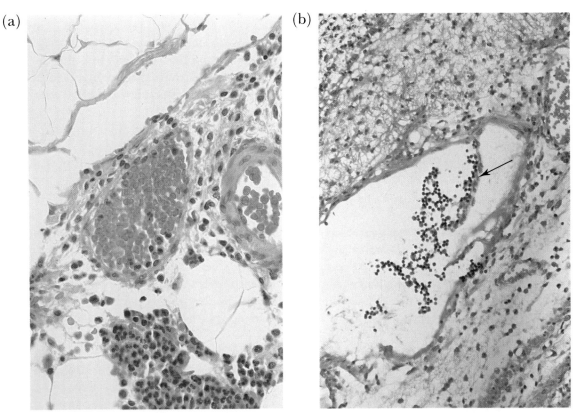

Fig. 15.1   Acute inflammation of the appendix. (a) Margination and the accumulation of polymorphs in the extracellular spaces. (× 400.) (b) Increased lymph flow in a much enlarged lymphatic vessel with valve (centre). (× 200.) H & E.

remove the cause. The third stage is the removal of damaged tissue and subsequent repair. Let us look at each of these in turn.

Mobilization of defences into the injured area

The earliest response of the body to injury occurs almost immediately: there is an increased blood flow to the affected area. This appears to be a response of local vasomotor nerves to substances liberated by damaged cells, since it does not take place in denervated tissue (Brown *at al.*, 1968).

   Within minutes or even seconds, depending on the severity and nature of the injury, the blood-vessels in the damaged area become permeable to proteins. Protein-rich fluid passes out into the tissue spaces, initially from the post-capillary venules, but also from capillaries themselves.

What effects would you expect this to have on fluid exchange between the capillaries and the extracellular spaces? (Note 15.A)

The local release of histamine is associated with this increase in permeability.

What cell type occurs frequently in the connective tissues and is known to be rich in histamine? (Note 15.B)

A group of polypetides known as kinins also appears to cause the passage of a protein-rich fluid out into the extracellular space.

Electron microscopy suggests that the endothelial cells lining the capillaries and post capillary venules retract slightly from each other, opening up small channels between them through which relatively large molecules, such as gamma globulins, can leak out into the tissue spaces (Majno and Palade, 1961). Thus many proteins including those involved in the production of fibrin and in immunity against foreign molecules reach the injured area very rapidly.

The gaps that open between adjacent endothelial cells are very small, and do not permit cells to pass out. Yet cells are needed in the damaged area, particularly the polymorpho-nuclear leucocytes, and these have to leave the blood-vessels without allowing erythrocytes to escape.

What are the particular features of neutrophil polymorphs which might be useful at a site of injury? (Note 15.C)

The polymorphs within the post-capillary venules cease to be carried in the flowing stream of blood, roll slowly down the endothelial lining and finally stick to it — processes known as margination (Fig. 15.1). Next, they insert themselves slowly between adjacent endothelial cells, the close attachment of their cell membranes to those of the blood-vessel wall acting as a seal, preventing the loss of erythrocytes. So they migrate through the wall into the tissue spaces outside, the endothelial cells coming together again as the polymorph moves out.

The signal that attracts the polymorphs is not certain. The presence of plasma proteins in the tissue spaces seems to stimulate migration, as does damaged tissue itself. Although neutrophil polymorphs are usually associated in our thinking with the presence of bacteria, infection is not necessary to provoke migration. Whatever the trigger, the initial effect seems to be on the cell membranes of the endothelial cells and the polymorphs, making them more sticky. Leucocytes stick to each other in clumps within the blood-vessels as well as to the walls, while other constituents of blood, notably platelets, also tend to stick to endothelium in damaged areas.

Initially, polymorphonuclear leucocytes form the majority of emigrating cells. Mono-cytes enter the tissue slowly at first, but form an increasing percentage of migrating cells after several hours. Some believe that the polymorphs and monocytes migrate at the same relative rate throughout the inflammatory reaction: the increasing percentage of mono-cytes in the inflamed tissue at later times would, in this view, be due to their longer survival and their ability to divide. Lymphocytes remain within the blood-vessels in the early stages of inflammation.

What cell is the equivalent of the monocyte in connective tissues? What is its principal function, and how does it differ from the polymorph? (Note 15.D)

This stage of mobilization produces three main changes in the damaged tissue — first, an increased blood flow; then a protein-rich extracellular fluid, with consequent effects on the equilibrium between extracellular fluid and blood; and thirdly large numbers of white blood cells, initially mainly neutrophil polymorphs, but later monocytes.

Let us assume that an area of dermis has become infected. What features would you expect to see on clinical examination of the inflamed area, as a result of the changes described above? (Note 15.E)

Restriction of the damage and removal of the cause

Some injuries, such as the majority of bruises, remain uninfected. If the skin or any other epithelial surface is damaged, however, the possibility of bacterial invasion is high.

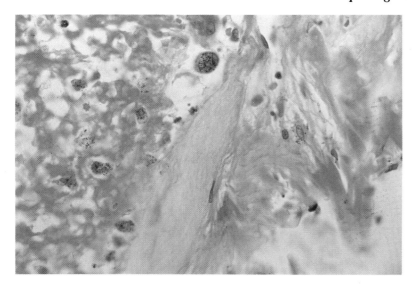

Fig. 15.2   Removal of fibrin and debris by macrophages; a few elongated nuclei of fibroblasts also present. H & E. (× 495.)

Whatever the cause of the injury, the restriction of its effects to a small, localized area is an obvious advantage to the body. Damage can spread by bacteria invading deeper into the tissues; damage can also be provoked by cell death locally. This releases enzymes and other active molecules into the extracellular spaces, with the possibility of damaging other cells directly. In addition, the osmotic effects of these molecules may slow the renewal of extracellular fluid, affecting the viability of cells indirectly. So cell death can be imagined setting up a chain reaction of further cell death and further release of damaging material.

Polymorphonuclear leucocytes are short-lived cells, equipped with pre-formed lysosomes and capable of phagocytosing bacteria in particular, but also cell debris, even under conditions of poor oxygenation. Their assault on invading micro-organisms is assisted by the presence of opsonins, a class of antibody molecule in the extracellular fluid. Neutrophil polymorphs cannot divide, so that all those in the neighbourhood of an infected area have arrived by migration from the blood. Macrophages, whether they were already in the tissues, or have arrived there as monocytes migrating from the capillaries, are particularly associated with the removal of cell debris (Fig. 15.2). They can divide, and can also synthesize new lysosomal material.

These two cell types between them are mainly responsible for the localization of damage, phagocytosing micro-organisms and breaking the cycle of cell death by removing cell debris.

At this stage of inflammation, antibody globulins clump and immobilize micro-organisms, making their phagocytosis easier. Plasma proteins in the tissue spaces may have other functions. Extravascular clotting is an important mechanism for filling the defects in tissues that follow cutting or tearing injuries, or grazes; it can also occur in other sites (Fig. 15.3).

What plasma proteins contribute to this mechanism? (Note 15.F)

Apart from local spread in the tissue spaces, two other possible routes for the spread of infection must be safeguarded: the lymphatics and the blood-vessels. Increased lymph flow from damaged areas starts as soon as leakage of plasma proteins occurs into the extracellular space, and this fluid is eventually returned to the circulating blood (Fig. 15.1b).

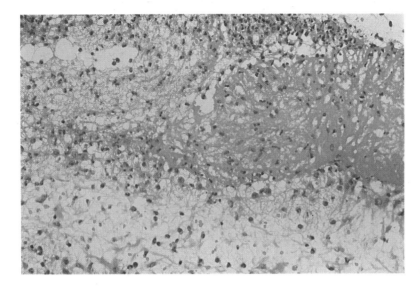

Fig. 15.3 Extravascular clot on the serosal surface of the appendix in acute inflammation: the original serosa ran across the centre of the field. H & E. (× 200.)

How does the body filter this lymph to prevent the spread of micro-organisms or of damage from dying cells? (Note 15.G)

Spread of infection or of damaged and dying cells by the blood stream from an injured area is largely prevented by intravascular clotting. Wherever the endothelium of a blood-vessel is damaged, platelets and the endothelial cells themselves become "sticky", so that plugs of platelets soon block smaller vessels, trapping red cells and providing a framework for the deposition of fibrin (Fig. 15.4).

Repairing the damage

Following, for instance, a scratch in the skin, with a temporary invasion by micro-organisms, the processes we have so far discussed usually dispose of the invaders. The gap in the epidermis remains, filled with fibrin clot in which are set remnants of the original

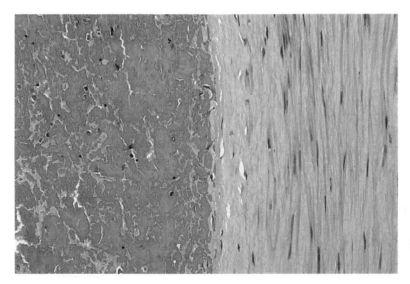

Fig. 15.4 Blood clot (left) inside a vein. H & E. (× 200.)

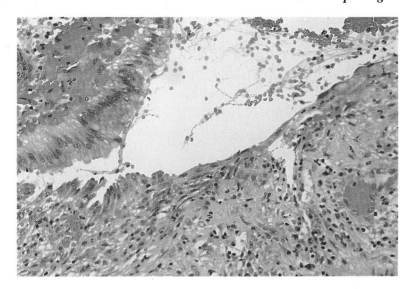

Fig. 15.5 Regrowth of epithelium at the edge of a gastric ulcer. Surviving columnar epithelium (left) continuous with a layer of flattened cells over connective tissue (centre). H & E. (× 190.)

tissues, together with debris including living and dead polymorphs. Blood-vessels at the injured site itself are closed by intravascular clot, but those immediately around the injured area are dilated. Two processes now go ahead together — the removal of debris, which may include removal of initially surviving tissue immediately round the damaged area, and its replacement by new tissue. While for convenience we will consider these two separately, both continue simultaneously: indeed, the division of the whole inflammatory process into three stages is slightly artificial, and one must expect to find the growth of new tissue, removal of invading bacteria and mobilization of polymorphs from the blood all occurring together at the one time.

The removal of debris. The macrophage, as we have seen, is the cell principally involved in the removal of debris. In addition, a further cell type begins to appear, the eosinophil polymorphonuclear leucocyte. This, like the neutrophil polymorph, is an end cell with a relatively short life, already equipped with primary lysosomes.
How does its function differ from that of the neutrophil? (Note 15.H)
Together these cells remove the debris and, in co-operation with the cells involved in repairing the damage, remodel the surviving tissue framework, mainly the collagen bundles. Recruitment of both macrophages and eosinophils from neighbouring blood-vessels is supplemented by cell division in the macrophage population.

Replacing damaged tissue. Reconstituting the damaged tissue involves several cell types. Epithelial cover is produced by an initial migration into the wound area of epithelial cells from the wound margins (Fig. 15.5). In the case of damage to epidermis, cells in the stratum basale and stratum spinosum in healthy epidermis around the damaged area change shape to become flattened, and migrate towards the gap in the epidermis, retaining contact with the healthy epithelial cells around the gap. The thin, translucent rim of migrating cells can often be seen with the naked eye, as it grows over the wound, eventually meeting in the centre of the wound to restore epithelial cover. The epithelial cells often grow across the wound below much of the clot and debris, which form a superficial, protective scab that finally separates as epithelial continuity is restored beneath it.

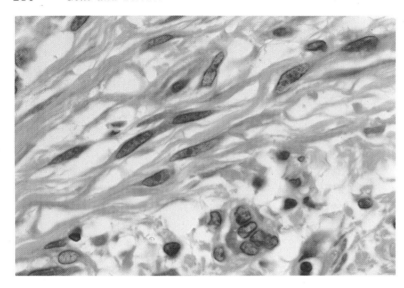

Fig. 15.6   Active fibroblasts in a healing surgical incision: a multinucleated phagocytic cell is centre bottom. H & E. (× 990.)

Migration is accompanied by an increased rate of cell division in the epidermis on either side of the wound. Once a complete layer of epithelial cells is established, differentiation begins, eventually restoring the keratinized, stratified squamous epithelium.

Beneath these epithelial changes, fibroblasts move in and start to lay down collagen, elastin and glycosaminoglycans as the fibrin clot and cellular debris is removed (Fig. 15.6). New capillaries grow into the area from the surrounding intact blood-vessels.

How does this process take place without permitting small leaks of blood cells and plasma into the tissues? (Note 15.1)

The speed and completeness of repair depend greatly on how much of the original tissue framework survives. A clean, surgical cut with the the edges held in contact with each other by sutures will rapidly return almost to the original pattern of organization of cells and extracellular fibres. A large, gaping wound fills first with clot, in which no collagenous framework exists: this is repaired with rather densely packed, new collagen with few blood-vessels and little of the precise yet flexible organization of normal dermis (Fig. 15.7). With the passage of several weeks, the new scar tissue contracts, drawing together the original wound edges and leaving a dense mat of collagen beneath the skin. Evidence is

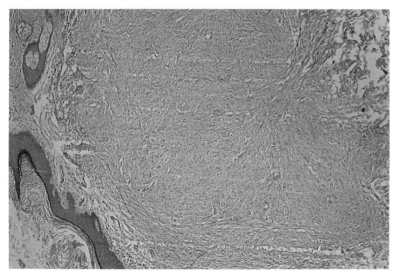

Fig. 15.7   Dense fibrous tissue of a dermal scar: epidermis bottom left. H & E. (× 40.)

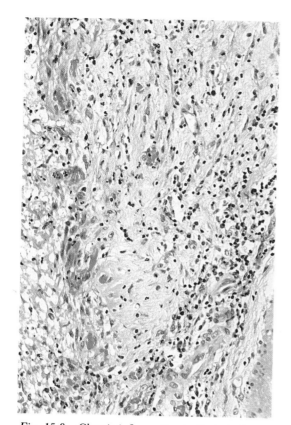
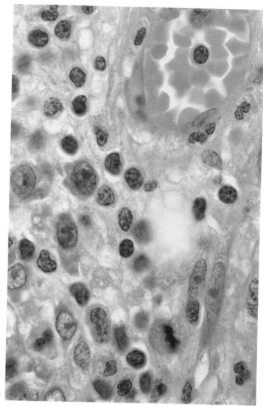

Fig. 15.8   Chronic inflammation in the bed of a gastric ulcer. (a) General view. (× 200.) (b) Lymphocytes and macrophages — the major cell types. (× 990.) H & E.

accumulating to show that the fibroblasts involved in repair synthesize actin and myosin in considerable quantities, and are capable of contraction: these cells are often called myofibroblasts. Their contraction is important in drawing together the margins of damaged areas (Gabbiani *et al.*, 1972). Repair is characterized by the presence of large macrophages, rather plump and active-looking fibroblasts, and increasing amounts of collagen (Fig. 15.6).

In addition to fibroblasts and their products, blood-vessels and nerves will grow into the injured area and lymphatic vessels appear again. As the need for them diminishes, macrophages wander off into the tissues around and travel up the lymphatics, to return to the pool of circulating monocytes.

Chronic inflammation

We have examined the sequence of events that follows a fairly trivial scratch in the skin. What happens if, for some reason, the cause of the injury persists for a longer time? This might happen if a thorn, instead of scratching the surface, remains embedded in the tissue, or if micro-organisms establish themselves and resist the body's attempts to kill and remove them. It also happens if local conditions prevent the normal processes of healing in some way: an acute injury to the epithelium of the stomach may not heal, since the fibrin clot covering the defect is not a competent protection against the very acid contents of the stomach. An uneasy equilibrium is then set up, which we call chronic inflammation (Fig. 15.8). The initial stages of exudation and mobilization of polymorphs are followed by

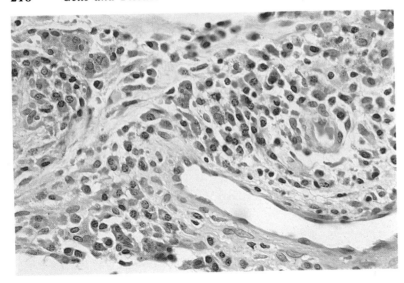

Fig. 15.9 A rheumatoid nodule from the capsule of a joint in rheumatoid arthritis. Many macrophages and plasma cells present. H & E. (× 400.)

attempts at healing, with the growth of new capillaries and much fibroblastic activity, but accompanied by continued damage. The cell population at such a site is rather variable, but includes a large number of macrophages and usually many lymphocytes. In time, the continuing attempts to lay down collagen given the tissues a scarred toughness which restricts the access of new blood-vessels and limits the exudation in response to fresh

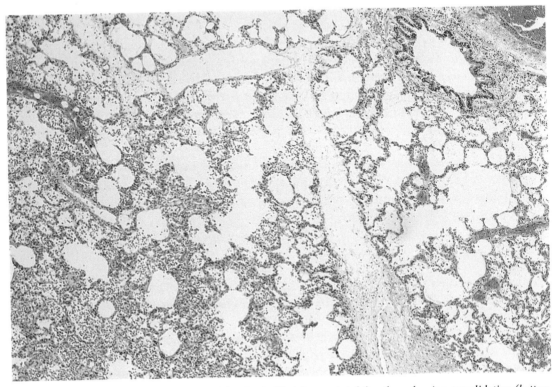

Fig. 15.10   Bronchopneumonia, with some alveoli air-filled (top and right), others showing consolidation (bottom left). H & E. (× 56.)

(a) (b)

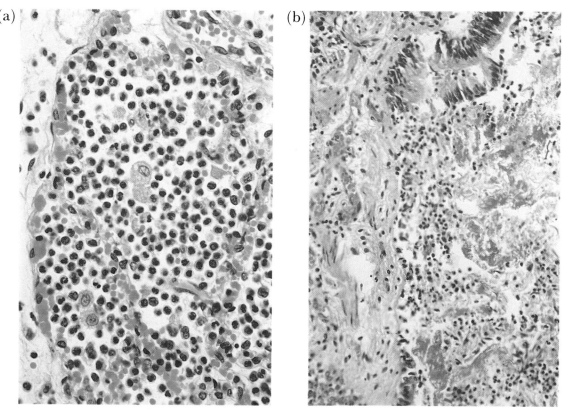

Fig. 15.11   Bronchopneumonia. (a) Alveoli filled with polymorphs, macrophages and fluid. (× 400.) (b) The wall of a small bronchus: the lumen (right) is filled with inflammatory exudate and the epithelium survives only as clumps of cells (above). (× 200.) H & E.

injury. If antigenic foreign material is present, plasma cells may be seen in large numbers: auto-immune disease, in which the body develops immunity against one or more of its own constituents, is associated with many plasma cells and eosinophils in the areas of chronic inflammation (Fig. 15.9).

Acute bronchopneumonia

Let us look now at acute inflammation in the lungs, as an example of the way in which the inflammatory process can be modified by the structural patterns in the tissue concerned. Figure 15.10 is human lung taken at post-mortem: compare it to Fig. 4.8.

The underlying pattern of alveoli, separated by thin walls of connective tissue carrying capillaries, can still be made out in places. The striking difference between this tissue and Fig. 4.8. is the replacement of clear air spaces by a rather structureless material in which many cells can be seen (Fig. 15.11a). This is protein-rich fluid, the inflammatory exudate, which has come from capillaries in the alveolar walls to fill the air spaces: histological fixation has precipitated the protein, which has stained faintly. Not all the air spaces are full of exudate: in places, clusters of clear spaces lie next to fluid-filled ones.

The cells in fluid-filled alveoli (Fig. 15.11a) include polymorphonuclear leucocytes and macrophages, and even occasional red blood cells. The capillaries of the alveolar walls look distended, and there may be evidence of margination.

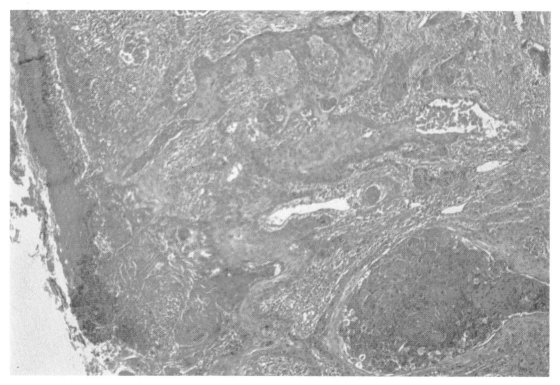

Fig. 15.12   Squamous cell carcinoma of the skin. Some recognizable epidermis survives (top left). H & E. (× 56.)

What is margination? (Note 15.J)

Sections through small bronchi sometimes are barely recognizable (Fig. 15.11b). The epithelium may be completely absent, or small tags only may survive; the underlying blood-vessels are grossly dilated, and there are many polymorphonuclear leucocytes and macrophages.

This is acute bronchopneumonia, in which infection of and damage to bronchi are followed by patches of acute inflammation in the substance of the lung. The fluid-filled alveoli are presumably ones supplied by small bronchi which are particularly severely affected.

Inflammation modified by the tissue. In the dermis, for instance, the inflammatory exudate remains in the extracellular spaces, which are relatively rich in collagen and GAGs, providing some limits to the amount of fluid that can leave the capillaries. Tissue pressure builds up to restrict the fluid entering the tissues, even closing capillaries. In the lungs, the epithelium is so delicate that the exudate enters the air spaces and fills them without opposing forces.

Bronchopneumonia can resolve with little structural damage: the mucosa can repair from surviving islands of epithelium. It is much more common, however, for scar tissue to result, with collagen deforming small bronchi and alveoli; if the periphery of the lung is involved, scar tissue may limit the free gliding of the pleural surfaces across each other. The result is a permanent reduction in the efficiency of the lung and an increased likelihood of further infection.

The histological picture and the patient. This, then, is the typical histological appearance of bronchopneumonia, in which infection, arriving down the bronchi, produces scattered patches of acute inflammation in the lung. From clues in the section and a very basic knowledge of how the body works, one can go on to make several predictions about the patient.

If many air spaces are filled with inflammatory exudate, the surface for oxygen exchange in the lung will be greatly reduced, and the normal balance between oxygen supply and rate of perfusion with blood will be altered. To counteract this, the patient will breathe faster, and his ability to move about will be limited by shortness of breath.

If acute imflammation is producing exudate in many bronchi and bronchioles, the body will attempt to rid itself of this material, which it does by coughing. The material coughed up will be the usual airway mucus, with the addition of the more watery exudate typical of acute inflammation. Polymorphs will be present in this material in large numbers, giving it the pearly look we associate with pus; there will often be red blood cells present as small haemorrhages occur from blood-vessels in bronchi from which the epithelium has been lost.

So we can predict that the patient is short of breath and breathing fast, has a bad cough, and is producing mucus with pus and perhaps some staining with blood.

Note that the processes of acute inflammation, which have evolved to protect and preserve the body, can be dangerous, even fatal, in some situations. The filling of lung alveoli with inflammatory exudate may kill by preventing oxygen exchange. The swelling that accompanies inflammation, if it occurs in the brain, may cause such a rise in pressure within the skull that vital functions of the brain may be interrupted.

Cancer

Cancer is an escape from the controls governing normal tissues. The rate of cell division in a group of cells is no longer linked to the needs of the tissue or the death rate in that subpopulation, but it continues unchecked. This is often associated with a failure of the newly produced cells to differentiate adequately. Cancer is a lay term covering a wide variety of conditions, however. Let us be more accurate, and use carcinoma, a malignant growth arising in an epithelial cell population.

Squamous cell carcinoma of skin

Figure 15.12 is a section of skin bearing a carcinoma. A small amount of normal epidermis can be seen at one edge of the picture. The abnormal cells occupy much of the rest of the slide. They retain many of the characteristics of an epithelial sheet. They are, nevertheless, clearly abnormal epithelial cells. Instead of the relatively thin, orderly epithelium seen at the section's margins, there is a massive accumulation of epithelial cells. The epithelium here has ceased to respond to the factors that control and organize it normally, resulting in an abnormal accumulation of cells.

Differentiation of cells is incomplete. Many of the new cells produced by mitosis make little recognizable effort to travel the path of differentiation seen in the normal epithelium, and the attempts to produce squames are so disorderly and lacking in polarity that they may produce small balls of squames deep in the tissue.

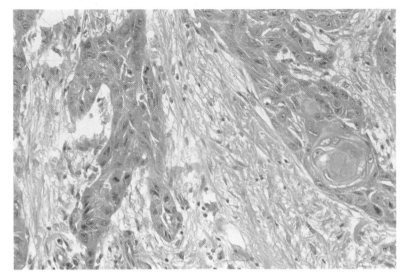

Fig. 15.13 Squamous cell carcinoma of the skin. H & E.
(a) Advancing edge of tumour, with cancer cells through the basal lamina, proliferating in the connective tissue spaces. (× 200.)

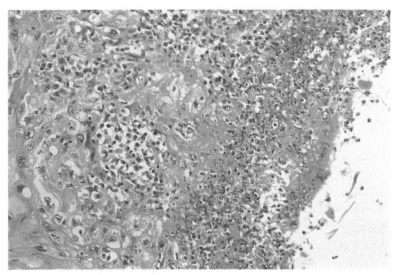

(b) Necrotic area at surface. (× 200.)

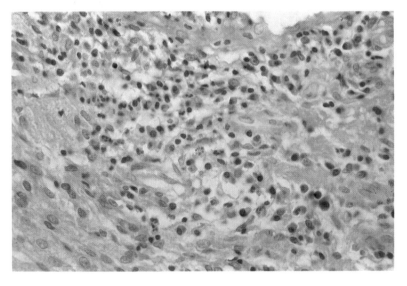

(c) Lymphocytic infiltration at edge of tumour. (× 405.)

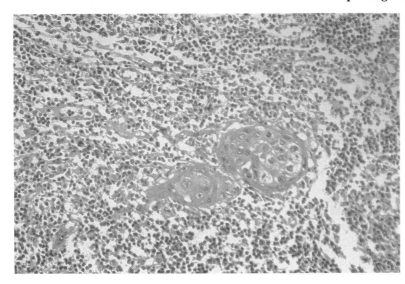

Fig. 15.14 Lymph node with secondary deposit (centre) from squamous cell carcinoma of skin. H & E. (× 200.)

The epithelial cells have ceased to respect the basal lamina. They are through this barrier and invading the connective tissues beneath in large numbers (Fig. 15.13a).

Finally, communication between cells in the epithelial sheet seems disturbed, to judge by the very variable appearance of adjacent cells.

Cells within an epithelial cancer have ceased to receive or respond to information in their environment, which normally produces growth limited to the immediate needs of the tissue, differentiation to a fully functional state, and the efficient and economical organization characteristic of healthy tissues.

How does the body respond to this growth of disorderly cells? The response we see is not just to the cells themselves, of course. With such a disorganized epithelial sheet, the surface layers are often too far from capillaries to remain alive, and a large patch of dying cells, presumably invaded by bacteria, can be seen at the surface of the growth (Fig. 15.13b). Around this, particularly near the edges of normal epithelium, there is a typical acute inflammatory reaction with many polymorphs. But though the cancer cells receive messages from surrounding cells poorly, if at all, they do signal their own needs. Cancers stimulate the growth of new capillaries from surrounding vessels by producing a compound called tumour angiogenesis factor. Without this support from capillaries and an appropriate connective tissue framework, cancers would never grow beyond a diameter of perhaps 1 mm: at this stage, cell death from anoxia and lack of nutrients would balance cell division at the periphery of the growth. The fact that these cells, while deaf to incoming messages, still signal to, and produce supporting responses from, surrounding tissues enables them to continue growng almost indefinitely.

At the deeper margins of the growth, a massive chronic inflammatory reaction can be seen with many lymphocytes, plasma cells and eosinophil polymorphs (Fig. 15.13c). The extent of this reaction varies with the cancer: some provoke it, others do so hardly at all. Cancers arise from cells of the host body, and many of the proteins and surface antigens they produce are indistinguishable from those of normal cells. They sometimes do produce antigenic material, either from genes normally held repressed or as a result of some change or mutation in their own genetic mechanism, and the presence of such antigens will provoke the chronic inflammatory response seen here. Such a response will swing the balance between cell death and cell division towards death, resulting in slower growth of the tumour. The presence of such an inflammatory response to a tumour is one of the clues looked for by the histopathologist, who has to predict the future behaviour of the cancer from its histological appearances.

Diagnosis of the abnormal rests firmly on knowledge of normal tissue organization. The same effort to relate patterns of structure to function, which has been emphasized in normal cells and tissues throughout this book, can produce a great deal of useful information when you are faced with abnormal patterns which are new to you. This is particularly true of cancer, essentially a disease of tissue organization.

The spread of cancer by metastasis

Figure 15.14 is a section of lymph node from a patient with a similar carcinoma to that shown in Figs 15.12 and 15.13. It illustrates one further important feature of carcinoma — the ability of abnormal cells to spread and to colonize distant tissues.
How do you think such cells might reach a lymph node? (Note 15.K)
In this particular site, it is difficult to recognize any inflammatory reaction to their presence, though sometimes areas of dying cells may be seen which contain many neutrophil polymorphs.

The way cancer cells lose touch with the original tumour and float off in lymph channels and even the blood stream tells us that they have lost some of the close cell-to-cell contacts typical of normal epithelia. Their ability to grow fast in quite strange environments is another aspect of their escape from the controls that regulate the organization of cells in normal tissues.

Summary

What has been learnt from this chapter? It certainly has not taught you histopathology, or the biology of cancer. I hope it has presented to you a view of tissues and organs as living, responding to damage and insults, changing from the patterns we accept as normal in attempts to maintain function and to repair themselves. By looking, however briefly, at a carcinoma of skin we see the potentially fatal results of failure in a control system — a system which we have hardly begun to understand, yet one the effects of which we see every time we note the remarkable order and precision of organized tissues.

This chapter should have achieved something else. The rest of the book has shown you how the patterns cells make can be interpreted in terms of function. This attempt to understand the significance of patterns can be applied just as successfully to disordered structure, and a relatively small number of principles can assist you to reason sensibly about conditions you have never seen before.

Further reading

Bartlett, G.L. (1979). Milestones in tumour immunology. *Semin. Oncol.* **6:4**, 515–525. This number of the journal contains several "milestone" articles on cancer.

Burnet, F.M. (1978). Cancer: somatic-genetic considerations. *Adv. Cancer Res* **28**, 1–29. An old dispute looked at by one of biology's great theoreticians.

Diegelmann, R.F. *et al.* (1981). The role of macrophages in wound repair. *Plast. Reconstruct. Surg.* **68**, 107–113. A short, recent summary.

Macleod, A.C. (1973). "Aspects of Acute Inflammation." Scope monographs: Upjohn Company. A clear and well-illustrated account of acute inflammation.

Nicolson, G.L. (1979). Cancer metastasis. *Scient. Am.* **240:3**, 50–60. Readable account.

Olsson, I. *et al.* (1980). The role of the human neutrophil in the inflammatory reaction. *Allergy* **35**, 1–13. Summarizes recent work.

Potter, V.R. (1980). Initiation and promotion in cancer formation: the importance of studies on intercellular communication. *Yale J. Biol. Med.* **53:5**, 367–384. Looks at the phenomenon of cancer in terms of cell communication.

Robbins, S.L. and Angell, M. (1976). "Basic Pathology." W.B. Saunders, Philadelphia. Probably the best undergraduate text in pathology. Try Chapters 2 and 3.

References

Brown, J.H., Kissel, J.W. and Lish, P.M. (1968). Studies on the acute inflammatory response. I. Involvement of the central nervous system in certain models of inflammation. *J. Pharmac. exp. Ther.* **160**, 231–242.

Gabbiani, G., Hirschel, B.J., Ryan, G.B., Statkov, P.R. and Majno, G. (1972). Granulation tissue as a contractile organ: a study of structure and function. *J. exp. Med.* **135**, 719–734.

Majno, G. and Palade, G.E. (1961). Studies on inflammation. I. The effect of histamine and serotonin on vascular permeability: an electron microscopic study. *J. biophys. biochem. Cytol.* **11**, 571–605.

16
What have we achieved?

Recurring patterns

Histology is largely made up of recurring patterns, many of which can be associated with particular functions. This book has presented many of these links between patterns and function, and has made them familiar to you. The standard textbook of histology, on the other hand, is a series of thorough descriptions of tissues and organs, arranged under organ systems. This design makes it difficult for the student starting histology to recognize the common patterns. They are nevertheless there, in the mass of data.

Patterns and their functional equivalents are clearly seen in electron micrographs of cells. Any micrograph is built up out of a limited range of recognizable organelles, each with a specific function. Their relative frequencies and their arrangement within the cell indicate the major task that it carries out. Cells that synthesize proteins for intermittent secretion have a characteristic and easily recognizable appearance, and can be distinguished from cells synthesizing proteins for their own growth, or those synthesizing steroid hormones, and so on.

Recurring patterns are just as obvious when one turns to the organization of cells into tissues. Epithelia can readily be recognized, whether they line free surfaces or glands, and the patterns of cells in an epithelium specialized for absorption from the lumen is quite different from that in an epithelium that protects underlying tissues from hypertonic urine or from abrasion and drying out by contact with air. Connective tissues, again, have relatively few different extracellular components. Their combination in the body can produce many different appearances, but, given a grasp of the functions of collagen, of cartilage matrix, of elastic fibres and so on, this range of apparently different connective tissues can be interpreted in terms of its characteristics in the living body.

At a higher level still, recurring patterns are found in the organization of tissues into organs. Tubes that propel their contents all have certain easily recognizable features which differentiate them from tubes that regulate flow but do not propel. Sites where particulate matter and soluble proteins are filtered from a body fluid have a common pattern.

Even the responses of the body to injury follow certain general patterns, which make sense if one recognizes the need to mobilize responding cells, to limit damage and to repair it.

Patterns and learning

What value is it to you to have looked at these patterns?

You certainly do not know all of histology and histopathology. But these patterns give you a framework of knowledge which allows you to look at a microscope slide or an electron micrograph with interest, and to make sensible predictions about functions from the patterns you see there. This framework will inform and guide all your future work in histology and histopathology, helping you to link the former with physiology and biochemistry, the latter with the disease process and the signs and symptoms of the patient. Your future work will reinforce and extend this framework of pattern recognition so that it becomes a familiar and almost unconscious part of your growing skills.

Histology is the meeting place of physiology and biochemistry with structure, with micro-anatomy. Unfortunately, many students find it a mass of fragmented facts, difficult to learn and remember and largely unrelated to "the real business of medicine". Given the framework of knowledge in this book, the facts become more reasonable and predictable, the descriptive lists in the histology textbooks start to make sense, and your intelligence can take much of the load off your unsupported memory. Once that begins to happen, the subject becomes more enjoyable, and a support to the associated topics of physiology, biochemistry and anatomy. You may even have time to wonder at the beautifully precise and disciplined order which the varied populations of cells show in building up and maintaining our bodies.

If I have directed your intelligence and your interest towards the study of cells and tissues, I have succeeded.

Notes

Chapter 2

2.A The projection lens lies just in front of the light source which is at or near its focal plane. The lens converts the filament of the lamp into an even source of light, which fills the lens (p. 6).

2.B There will be an increased chance of stray light, or glare (p. 6).

2.C The light beam illuminating the specimen will have a larger solid angle than is needed to match the NA of the objective lens. This produces glare (more accurately, loss of image contrast) and slight image degradation from rays passing through the very edges of the objective lens (p. 6).

2.D The diameters of the apertures in the two iris diaphragms should be matched to the objective lens, and will need adjusting (p. 6).

2.E Every interface between media of different refractive index produces some glare and image degradation (p. 7).

2.F See Fig. 2.16!

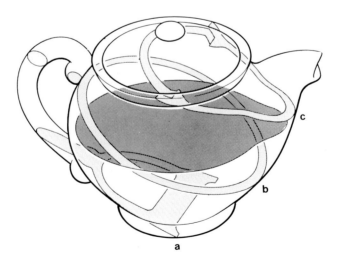

Fig. 2.16   A drawing of a half-filled teapot, from which the sections in Fig. 2.11 were taken.

2.G The light source is situated so that the direct light beam cannot enter the objective lens. Only light that is scattered or reflected can enter and contribute to the formation of the image (p. 14).

2.H The limit of resolution of the light microscope is about 0·2 μm. The limiting factor is the wave motion of the photons (p. 4).

Chapter 3

3.A The nucleus is large (compare with the mitochondria) and has little condensed chromatin: many genes are being transcribed. There is a large nucleolus, indicating synthesis of new ribosomes.

3.B This nucleus would appear large, rounded and pale staining, and the nucleolus would be obvious and darkly stained. By contrast, the plasma cell (Fig. 3.6) would have a smaller nucleus, with densely stained chromatin packed around the nuclear membrane and extending into the centre of the nucleus, giving an appearance often referred to as "cartwheel".

3.C There are 1 000 000 nm in 1 mm. A structure 10 nm in diameter magnified 65 000 times would be 0·65 mm across. Subunits cannot be seen in Fig. 3.8 because lack of contrast prevents us from distinguishing them. Ribosomes (15–20 nm) can be seen, as they are more heavily stained than surrounding structures.

3.D The resolving power of the light microscope is about 0·2 μm, the width of a mitochondrion (p. 4). Under the best conditions, mitochondria can be seen with the light microscope, particularly if they are arranged in parallel series, as in Fig. 3.9.

3.E A large nucleus with relatively little condensed chromatin, containing one or more prominent nucleoli.

3.F Such enzymes will be synthesized in RER (p. 30).

3.G The digestive enzymes will be transported first to the convex, receiving face of the Golgi apparatus, in small vesicles surrounded by inner membrane. The enzymes will emerge from the concave, mature face of the Golgi packaged in vesicles of outer membrane, and will be transported in these vesicles to the phagocytic vacuoles (pp. 30–32).

3.H Phase contrast and interference contrast are the two techniques most widely used to examine living cells (p. 13).

3.I Basophilia is a property of acidic compounds, which bind basic dyes during staining. Of the acidic compounds in the cytoplasm, ribosomal RNA is the principal one that survives fixation and histological processing. Cytoplasmic basophilia suggests the presence of many ribosomes, and therefore active protein synthesis (p. 10).

3.J A lysosome is a membrane-bound cytoplasmic organelle containing catalytic enzymes, in particular acid phosphatase. A phagocytic vacuole contains extracellular material which the cell has ingested. It becomes a secondary lysosome when primary lysosomes fuse with it, pouring their digestive enzymes into it. Secondary lysosomes vary widely in appearance, reflecting differences in the material ingested and in the time over which digestion has proceeded (pp. 34, 35).

3.K Figure 3.18 shows a small, round cell. Its nucleus is small and almost filled with condensed chromatin, suggesting that very few genes are active. The cell membrane is smooth, implying that there is little exchange going on with the surrounding fluid. The cytoplasm contains very few organelles, apart from a few mitochondria. Numerous free ribosomes can be seen, often in clusters. The major activity of this cell is the synthesis of protein which is retained within the cytoplasm. This is an erythroblast, a precursor of the red blood cells: it is synthesizing haemoglobin, and doing very little else. With the light microscope, it would appear as a small, rounded cell with a small, darkly stained nucleus. Its cytoplasm would be basophilic from the large number of ribosomes, perhaps modified slightly by the red colour of the haemoglobin.

Chapter 4

4.A The plane of fracture separates the two lipid layers in the membrane (p. 23). The technique allows one to examine large areas of membrane *en face*. IMPs are shown as raised dots on the membrane surface.

4.B The diameter of a red blood cell in fixed tissue is about 7·2 μm. Magnified × 300, this is just over 2 mm.

4.C Protein synthesis for export is associated with large amounts of RER in the cytoplasm, and a prominent Golgi apparatus. Intermittent secretion implies the presence of storage granules, probably clustered in the apical cytoplasm. By light microscopy, the basal cytoplasm is basophilic, suggesting a high content of RNA, and the apical secretion granules are clearly visible (p. 32).

4.D The mitotic rate is low in pancreatic epithelium. Mitoses, when they do occur, take place in the epithelium itself, without any structural arrangement to shield them from the lumen (p. 46).

4.E One might expect an increased surface area for mucus production in glands opening on to the surface (p. 46).

4.F Gap junctions (p. 51).

4.G Mitochondria (Fig. 4.14).

4.H These are stem cells capable of dividing, and of differentiating into either goblet or ciliated cells (p. 46). (Note that recent work suggests that some mucous cells may also be capable of dividing.)

4.I There is a steady rate of cell division in the epithelium of the bladder, similar to that in the skin. Dividing cells lie on the basal lamina, in the most sheltered and controlled micro-environment available (p. 46).

4.J One might expect mucous glands in the wall of the oesophagus to open on to its surface (p. 46).

Chapter 5

5.A A basal lamina contains fine, collagenous fibres set in a matrix of GAGs. Suggested functions are physical support for the overlying epithelium or endothelium, the prevention of contact between cell groups on either side of the lamina and the prevention of diffusion of some molecules across the lamina, while permitting water and some solutes to pass (p. 44).

5.B Such a cell will have many, scattered areas of RER, often in the form of distended sacs, a prominent Golgi apparatus, many mitochondria, and a few vacuoles, often associated with the surface membrane. There will be no secretion granules (p. 37 and Fig. 3.16).

5.C Smooth ER and the Golgi apparatus (pp. 30–32). Remember that these are polysaccharide-containing molecules destined for export from the cell.

5.D A macrophage is actively synthesizing lysosomal enzymes. It will have a nucleus with a prominent nucleolus and relatively little condensed chromatin. The cytoplasm will contain many short runs of RER, a prominent Golgi apparatus, and many small, primary lysosomes. Phagocytic vacuoles of varying size, shape and contents may be present, together with secondary lysosomes and residual bodies (p. 39 and Fig. 3.17).

5.E Standard histological techniques will dissolve out the fat, leaving the thin cytoplasmic rim and flattened nucleus around a central space (p. 9 and Fig. 5.6).

5.F No. Lipid is already "separate" from the aqueous cytoplasm by being in a hydrophobic droplet. It makes no sense to isolate it further with a phospholipid membrane (p. 36).

5.G Metachromasia occurs when a stain, associated with a particular tissue component, takes on a colour other than its usual one (p. 3).

5.H At 1·5 nm, magnified 95 000 times, we have 1 425 000 nm or just under 1·5 mm: 1·5 nm is resolvable by TEM under ideal conditions, but no specific stain is available to give adequate contrast (p. 19).

5.I Verhoeff's stain will demonstrate elastic fibres (p. 10). They appear black. Note that other stains for elastin also exist, which you have not yet met.

5.J The "collagen fibres" seen with the light microscope are not single fibres, but bundles of fibres, scores or even hundreds of them, running parallel to each other (p. 73).

Chapter 6

6.A This is a continuous layer consisting entirely of cells. It is an epithelium (p. 46).

6.B These are cut surfaces.

6.C The tissue is not homogeneous. Beneath the epithelium is a dense layer, probably collagen. Below this again is a wide zone of net-like holes, probably fat cells. A few groups of cells which could be epithelial can be seen in the fat.

6.D The layer presumed to be collagen is dense, and not all the fibres run in the same direction. The fat cells have strands of collagen running through, some carrying tubes which could be blood-vessels. Apart from the clusters of cells which could be epithelial, no other specialized cell groups are obvious.

6.E This is a stratified, squamous epithelium with keratinization of the surface cells. The outer environment is air, and the keratin is waterproof, preventing fluid loss (p. 61).

6.F A tubular, coiled, simple gland (pp. 46–48).

6.G Lipids dissolved out in processing (p. 9).

6.H Holocrine (p. 47).

6.I These cells divide to replace surface cells lost through abrasion (p. 61).

6.J Desmosomes with microfilaments converging on them in the cytoplasm (p. 50).

6.K Collagen resists stretching. If arranged in a random, felt-like layer, some movement is possible in any direction until enough fibres are lined up to oppose further movement (p. 72).

6.L Elastic fibres, assisted by GAGs (p. 75).

6.M Strands of collagen, linking the dermis to the underlying tissues.

6.N When hot, the skin is flushed, with high blood flow. When cold, it is pale, with restricted blood flow.

6.O Such a cell might pump ions and water into the canaliculi (p. 29).

6.P The intermittent secretion of a material that includes protein and carbohydrate (pp. 30–32).

6.Q A large nucleus with little condensed chromatin. Abundant SER and a prominent Golgi apparatus. A moderate number of mitochondria. Lipid droplets accumulating in the cytoplasm as synthesis proceeds (p. 30).

6.R Holocrine secretion involves the death and disintegration of the cells, the entire contents contributing to the secretion (p. 47).

Chapter 7

7.A This process is summarized in Fig. 5.1.

7.B Little of the genome is being transcribed, and there is very little formation of new ribosomes going on (p. 27).

7.C Secretory granules and primary lysosomes are manufactured on the RER and processed through the Golgi apparatus. The lack of these organelles suggests that synthesis of protein and its packaging in external membrane have ceased.

7.D About 7·2 μm in fixed material (p. 3).

7.E The membrane-bound granules with a positive reaction for acid phosphatases are primary lysosomes. The enzymes present in these are synthesized on the RER. The presence of many free ribosomes in the cytoplasm suggests the synthesis of structural proteins for use within the cell, in other words a cell that is still growing (pp. 34, 29).

7.F Large, phagocytic cells in connective tissue are called macrophages (p. 39).

7.G By freeze fracture, tight junctions look like meandering lines of intramembranous particles, joining and dividing again to enclose irregular spaces (Fig. 4.3). At tight junctions, the membranes of adjacent cells are fused together along the lines of IMPs obliterating the extracellular space and sealing off the lumen of the capillary from the surrounding extracellular space (p. 49).

7.H Pinocytotic vesicles are small droplets of fluid, without identifiable solid material in them, within the cytoplasm of cells. They are formed by an invagination of the cell membrane, producing a flask-shaped structure whose neck pinches off from the surface membrane. Such vesicles may travel across a cell to fuse with the opposite cell membrane, and liberate their contents again (p. 33).

Chapter 8

8.A The osmotic pressure exerted by the plasma proteins is important in drawing extracellular fluid back into the venous ends of capillaries: this can only happen in the absence of significant levels of protein in the extracellular fluid (p. 65).

8.B The fluid will flow more slowly.

8.C Reticular fibres are fine collagen fibres (p. 75).

8.D The eosinophil polymorphonuclear leucocyte (p. 94).

8.E Lymphocytes are small, spherical cells with a diameter about 7 μm, containing a densely stained nucleus and having only a small rim of basophilic cytoplasm (p. 96).

8.F Radioactive isotopes have been widely used in labelling and tracing experiments of all sorts (p. 17). The DNA of cells is the component with the slowest turn-over rate, and hence the best molecule into which to introduce radioactivity for cell tracing experiments.

8.G The nucleus will get larger, with less condensed chromatin, as more genes are activated. The nucleolus will become larger and more prominent, as the synthesis of new ribosomes proceeds. The cytoplasm may appear basophilic, with many free ribosomes. The cell as a whole will become larger.

8.H A cell synthesizing a single protein is using relatively little of its genome. The nucleus will be relatively small, with a lot of condensed chromatin. The cytoplasm will be basophilic, because of the high content of RER. Since secretion is not intermittent but continuous, there will not be secretion granules (pp. 27, 32).

8.I The animal would never develop the ability to mount an immune reaction against cell-bound antigens.

8.J The vessels within the lymphoid tissue have a total, cross-sectional diameter far greater than that of the vessels leading to the lymphoid tissue. In addition, the vessels within the lymphoid tissue have many fine strands of collagen fibres running across the lumen, producing turbulence (p. 107).

8.K No: in the absence of antigens, there is no stimulus for the expansion of subpopulations of lymphocytes by cell division.

Chapter 9

9.A The techniques of phase contrast and interference contrast microscopy allow living, unstained cells to be observed with the light microscope (p. 13). The requirements of having the specimen in a vacuum and the high dose rate of electrons prevent living cells from being observed by electron microscopy (p. 19).

9.B Desmosomes and the intermediate junction (or zonula adherens) have this structure (p. 50).

9.C The presence of ATP as a source of energy, and the presence of Ca^{2+} ions which are needed for the attachment of myosin to actin (p. 117).

9.D At a desmosome, the membranes of adjacent cells lie parallel and about 20 nm apart. One or more lines of electron-opaque material lie in the gap, parallel to the membranes. There is an obvious line of dense material on the cytoplasmic side of the membranes, and many microfilaments radiate out into the cytoplasm from this dense material. The attachment sites in smooth muscle lack the structures lying between the membranes (p. 50).

9.E In sectioned material, gap junctions are places where the membranes of adjacent cells lie very close to each other — about 2 nm apart. By freeze fracture they are seen to be regular arrays of intramembranous particles (p. 51).

9.F Gap junctions (p. 51).

9.G Van Gieson's stain colours cytoplasm a greenish-yellow, and collagen red. Various trichrome stains also differentiate between cells and collagen (p. 10).

9.H Regularly arranged thick filaments, with the thinner actin filaments lying between and around them.

9.I The A-band (the myosin filaments) will remain the same length, but the I-band (that part of the actin filaments beyond the ends of the myosin) will shorten. At the same time, the H-band (the central part of the myosin filaments, without actin around them) will shorten as the central ends of the actin filaments from opposite ends of the sarcomere are drawn closer together.

9.J The sarcolemma (p. 123).

9.K The membranes of adjacent cells run parallel, and about 20 nm apart. No structure is visible between the membranes at this site, but there is a dense thickening of the cytoplasmic side of the membranes and many microfilaments radiate out into the cytoplasm from that thickening (p. 50).

Chapter 10

10.A Collagen, elastic fibres, GAGs and extracellular fluid (p. 72).

10.B Glycosaminoglycans (GAGs) are large, roughly spherical molecules containing aminosugars. They have many charged groups exposed on their surfaces, binding cations and thus holding water relatively tightly within a given radius, or domain (p. 76).

10.C The matrix of cartilage is basophilic, staining blue with haematoxylin (p. 10).

10.D With the TEM, abundant RER is obvious in the cytoplasm, together with a Golgi apparatus that is clearly active. Since secretion is continuous, there are no secretion granules. The nucleus contains little condensed chromatin (p. 38).

10.E Fibro-cartilage would resist tension more strongly than hyaline cartilage, due to its higher content of collagen fibres.

10.F As movement proceeds, collagen fibres which were originally oriented in many directions become increasingly aligned to resist further displacement. Bands of collagen accompany blood-vessels across the plane that permits independent movement (p. 86).

10.G Adjacent bone surfaces would be joined by very short but strong bands of collagen to restrict movement.

10.H Cartilage. This joint consists of a central layer of fibrocartilage, between two layers of hyaline cartilage on the bony surfaces.

10.I A fluid-filled space, from which cells, fibres and capillaries are excluded (p. 148).

10.J Hyaline cartilage.

10.K The synthetic cell would have RER in the cytoplasm and a prominent Golgi apparatus but no secretory granules. The phagocytic cell could have RER, Golgi apparatus and primary lysosomes, together with phagocytic vacuoles and residual bodies.

10.L Fenestrations are a specialization of the walls of capillaries to assist transfer of material between blood and extracellular fluid (p. 100).

10.M The force compresses the proteoglycan molecules. On removing the deforming force, internal repulsion between the negative charges along the molecules restores the proteoglycans to their original shape. Collagen running through the cartilage limits and modifies the changes in shape produced by compressing the proteoglycans (p. 136).

10.N In addition to the deformation produced by forces of short duration (Note 10.M), continued loading drives water out of the compressed area, water that was previously held bound by proteoglycans (p. 137).

10.O Such a capsule should be made of bundles of collagen fibres, oriented to be fully extended at the desired limit of movement.

Chapter 11

11.A First, look naked eye: then, with a hand lens. Next, go to the low power of the microscope, identifying the major tissues and features present. The higher powers of the microscope should be used only to answer specific questions that emerge from your low-power observations (p. 78).

11.B Striated muscle fibres are large, multinucleate cylinders, with the nuclei lying peripherally beneath the cell membrane; the contractile elements are organized in precise and orderly fashion, producing cross-striations. Cardiac muscle is similar in the organization of contractile elements but has smaller fibres, with one or two nuclei, centrally placed, and very obvious junctional regions where one fibre is attached to the next. Smooth muscle fibres are single cells, tapering with central, elongated nuclei and irregularly arranged contractile elements (pp. 122, 128, 131).

11.C Smooth muscle contracts slowly, fatigues slowly and relaxes slowly: it is remarkably economical. Striated muscle contracts fast, fatigues fast and relaxes fast: it is used for rapid, precise and powerful movements. Cardiac muscle combines the rapid responses of striated muscle with inherent rhythmicity (pp. 122, 127, 131).

11.D Veins resemble arteries in having an intima, a media and an adventitia. The muscle coat is much thinner, since contraction does not have to take place against significant pressure in the lumen. Since there is no pulsatile pressure in veins, they do not have the elaborate organization of elastic fibres seen in arteries.

11.E A simple epithelium suggests that exchange of some sort is taking place between the luminal contents and the underlying tissues. A multi-layered epithelium suggests protection against exchange (p. 52).

11.F Such sacs are lined by flattened, mesothelial cells, and contain a thin film of fluid rich in glycosaminoglycans, from which cells, fibres and blood-vessels are excluded (p. 148). Examples include the peritoneal, pleural and pericardial sacs and tendon sheaths.

11.G This layer is a serous membrane. The cells are mesothelial that line it. It is called peritoneum in this site. It lines the coelomic cavity.

11.H The presence of rigid walls, preventing collapse, suggests that air occupies the lumen (p. 158). This occurs in the respiratory tract.

11.I Its wall does not contain muscle fibres, since it neither propels the contents nor regulates flow. The duct is surrounded by collagen, to prevent it from ballooning when secretion increases the intraluminal pressure; there may also be some elastic fibres.

11.J Exchange takes place between the luminal contents and the tissue, so a simple epithelium is likely. Blood capillaries below the epithelium are the most probable source of water and bicarbonate ions.

11.K A low columnar or cuboidal epithelium is likely, without any notable specialization at the light microscopic level.

11.L The tube will have an adventitia, since large, sliding movements relative to surrounding structures do not occur. Lymphoid aggregates will be rare, and so will mitoses.

11.M Cilia are membrane-covered extensions of the cell, containing microtubules arranged in a characteristic "9 + 2" pattern (p. 58 and Figs 4.13, 4.14). Microvilli suggest absorption of materials into the cell (p. 56).

Chapter 12

12.A The glycocalyx consists of polysaccharide molecules attached to the outer surface of the cell membrane, forming a layer of branching chains directed into the extracellular space (p. 24).

12.B Epithelia (p. 51), smooth and cardiac muscle cells (pp. 122, 131).

12.C Glycosaminoglycans are present in basal laminae (p. 44).

12.D Abundant SER, and the absence of secretion granules (p. 30). The mitochondria often have characteristically tubular cristae.

12.E Fenestrae are small pores in the wall of the capillary, at which the two opposing cell membranes of the endothelial cell come together to form a single diaphragm. Fenestration is thought to assist the passage of molecules into or out of the blood stream (p. 100).

12.F Phagocytosis by macrophages in the extracellular spaces, and the passage of extracellular fluid up lymphatic vessels, which takes the proteins past macrophages in the lymph nodes; these effectively filter the lymph before returning the proteins that remain in solution to the blood stream. Proteins accumulating in the extracellular fluid would exert an osmotic pressure opposing the return of fluid to the venous ends of capillaries (p. 66).

12.G Gated sodium and potassium channels (p. 173).

12.H In the central nervous system, non-myelinated nerves run through the cytoplasm of oligodendrocytes: myelinated nerves have a myelin sheath formed by them (pp. 174–176).

12.I Myelin, having a high lipid content, is dissolved out of the tissue in histological preparation, unless special steps are taken to preserve it (p. 9).

Chapter 13

13.A The nuclear membrane defines the nucleus, the environment in which the genetic material lies, and regulates the entry into it of substances from the cytoplasm. It thus protects the genetic material from cytoplasmic changes that might affect its translation, apart from those specific messages whose entry to the nucleus is permitted (p. 25).

13.B A circular ring of actin filaments just below the membrane contracts, progressively deepening the cleavage furrow (p. 119).

13.C Autoradiography is the use of photographic emulsions to detect the sites at which radioactivity is present in a specimen. Histochemistry demonstrates the distribution of a particular compound in the specimen. Autoradiography following the injection of a labelled precursor would only show the distribution of the newly synthesized fraction of the compound concerned: those fractions synthesized before the injection would be non-radioactive, and hence not demonstrated (pp. 17, 14).

13.D Since the cell cycle time is approximately 12 hours, cells in the crypts will have divided again at least twice, halving their content of radioactive DNA with each division.

13.E The monocyte leaves the blood stream to become a macrophage in the tissues. Free macrophages may become fixed macrophages in lymphoid tissue and elsewhere (p. 70). The circulation of lymphocytes is summarized in Fig. 8.8, and p. 112.

13.F Glial cells are the non-neuronal cells of the central nervous system. Oligodendrocytes are the cells which lay down myelin, the equivalent of the Schwann cells of the peripheral nervous system. Astrocytes are non-excitable cells capable of removing excessive potassium from the tiny volumes of extracellular fluid around neurones. Microglia appear to be phagocytic (pp. 181, 182).

13.G The final stage of DNA synthesis in the production of a sperm is that which precedes the first meiotic prophase (p. 194).

Chapter 14

14.A A large tube suggests transport of luminal contents. The two muscle coats imply that the tube propels the contents down the lumen (pp. 153, 156).

14.B A simple epithelium lining a tube suggests that exchange takes place between the luminal contents and the surrounding tissues (p. 52).

14.C The description of the nucleus suggests a cell that is active, with a considerable fraction of its genome in use. The principal function of the cell is not the synthesis of proteins, since there is not much cytoplasmic RNA (giving basophilia), nor are there secretion granules. The brush border consists of packed microvilli: this appearance suggests absorption from the lumen into the cell. The absence of basal striations (caused by infoldings of basal cell membrane with mitochondria lying between them) rules out ion pumping at the cell base as a likely function. This is the typical appearance of an absorptive cell (p. 56).

14.D Cell production must be balanced by cell death, to maintain normal structure. The cells produced in the crypts migrate up the surfaces of the villi to be shed from the tips of the villi. Here, irregularities of the epithelium can often be seen, with small tags of epithelial cells projecting into the lumen.

14.E The appearance of many lymphoid cells and nodules of lymphoid tissue beneath an epithelium suggests a high probability of invasion by micro-organisms from the lumen (p. 160).

14.F Terminal lymphatic capillaries have gaps in the basal lamina: the endothelial cells, which are tethered to surrounding connective tissue by delicate fibres, overlap each other if the tissue is compressed, but gaps open between them if the surrounding structures are spread apart by an increase in volume of the extracellular fluid. They thus act as a valve, permitting proteins and particulate material to enter the lymphatic when extracellular fluid volume is high, but preventing leakage out of the lymphatic when the tissue is compressed (p. 66).

14.G An exocrine gland secretes on to the external surface of the body, which includes the lining of the gut, respiratory tract and so on. A compound gland has a duct which branches. The term alveolar indicates that the secretory portion of the gland consists of a rounded group of cells clustered about a central lumen (pp. 46, 47).

14.H The outer surface of the duodenum has an adventitia, not a serosa.

14.I They appear to be coiled, tubular glands. The epithelium has a single cell type, with a densely stained, small nucleus at the base of the cell, often appearing rather flattened. The cytoplasm is large, homogeneous and faintly stained. This appearance is suggestive of a mucus-producing cell.

Chapter 15

15.A The renewal of extracellular fluid normally depends in part on the plasma proteins remaining within the capillaries. Once proteins in solution accumulate in the extracellular spaces, fluid does not return freely to the venous end of capillaries and to venules. The result is retention of fluid in the tissue concerned, with swelling (p. 65).

15.B The mast cell is rich in histamine, which is stored in granules in the cytoplasm (p. 71).

15.C Neutrophil polymorphonuclear leucocytes are fully differentiated cells specialized for phago-cytosis and the destruction of micro-organisms: they contain granules rich in anti-bacterial substances, and large, primary lysosomes. They can operate in relatively anaerobic conditions. They do not divide. They are capable of immediate attack on invading micro-organisms, without requiring time to synthesize the means to respond (p. 94).

15.D The macrophage is the tissue equivalent of the circulating monocyte. The macrophage lives much longer than the polymorph, and is capable both of dividing and synthesizing new lysosomes. It is essentially phagocytic, with a particular liking for cellular debris. It normally contains primary lysosomes, even as a circulating monocyte (pp. 70, 97, 98).

15.E Increased blood flow makes the area red and increases the local skin temperature. Increased extracellular fluid following escape of protein from blood-vessels produces swelling. Swelling and tissue damage produce pain.

15.F Prothrombin and fibrinogen are plasma proteins associated with clotting. Prothrombin, on contact with a substance called thromboplastin released from damaged tissues, becomes converted to thrombin. Thrombin, in its turn, is an enzyme catalysing the conversion of fibrinogen to insoluble fibrin. Fibrin forms a feltwork of intertwined fibrils, providing the basis for the clot. If platelets are liberated into the wound by haemorrhage, they also release a thromboplastin, as well as contributing both to the substance of the clot and to clot retraction (p. 102).

15.G Lymph enters the subcapsular sinuses of lymph nodes, and flows slowly down irregular channels or sinusoids, lined with highly phagocytic cells. These remove from the lymph all particulate debris and also micro-organisms, so that the lymph leaving a node is normally clear of such material (p. 106).

15.H The large, eosinophilic granules that give these cells their name contain preformed lysosomal enzymes. Unlike the granules of the neutrophil, they do not appear to contain specific anti-bacterial substances. Eosinophils are thought to be particularly involved in the phagocytosis and removal of antigen–antibody complexes (p. 94).

15.I An endothelial cell migrates out of the capillary wall, the neighbouring endothelial cells meeting to close the deficit as the first cell moves. Once lying between intact endothelium and basal lamina, the first cell divides. By this process of migration and division, small heaps of new endothelial cells transform into solid rods of cells, growing out from the wall of the original capillary. Only when one such rod meets and fuses with another similar rod growing in from elsewhere does the lumen open up, communicating with the capillaries at either end (p. 103).

15.J Margination is the attachment of nucleated blood cells to the luminal surface of the endothelium of a capillary or venule. It is the first step in the migration of such cells out of the blood into the tissues (p. 212).

15.K Small clumps or even individual cells loose contact with the rest of the epithelial sheet where it invades the connective tissues, and fail to form attachments to the cells and fibres of their new surroundings. Some of these cells get carried in the flow of extracellular fluid into the lymphatics and up to the regional lymph node. If they remain intact, the macrophages there accept them as part of the body, since they often do not carry foreign antigens on their surfaces.

Index